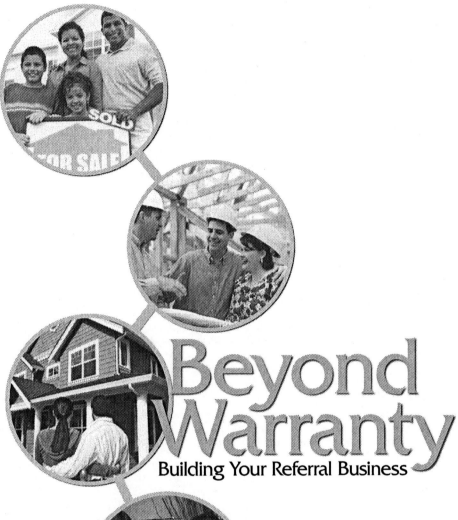

Beyond Warranty

Building Your Referral Business

CAROL SMITH

BuilderBooks.com®
A Service of NAHB
NATIONAL ASSOCIATION
OF HOME BUILDERS

BOOKS THAT BUILD YOUR BUSINESS

Beyond Warranty: Building Your Referral Business
Carol Smith

BuilderBooks.com®, a Service of the National Association of Home Builders

COURTENAY S. BROWN — Director, Book Publishing
DORIS M. TENNYSON — Senior Editor
NATALIE HOLMES — Book Editor
TORRIE SINGLETARY — Production Editor
CIRCLE GRAPHICS — Composition and Cover Design
MIDLAND INFORMATION RESOURCES — Printing

Gerald M. Howard — NAHB Executive Vice President and CEO
Mark Pursell — NAHB Senior Staff Vice President, Marketing & Sales Group
Lakisha Campbell — NAHB Staff Vice President, Publication & Affinity Programs

Disclaimer

This publication provides accurate information on the subject matter covered. The publisher is selling it with the understanding that the publisher is not providing legal, accounting, or other professional service. If you need legal advice or other expert assistance, you should obtain the services of a qualified professional experienced in the subject matter involved. Reference herein to any specific commercial products, process, or service by trade name, trademark, manufacturer, or otherwise, does not necessarily constitute or imply its endorsement, recommendation, or favored status by the National Association of Home Builders. The views and opinions of the author expressed in this publication do not necessarily state or reflect those of the National Association of Home Builders, and they shall not be used to advertise or endorse a product.

Printed in the United States of America

12 11 10 09 08   1 2 3 4 5

Library of Congress Cataloging-in-Publication Data

Smith, Carol, 1946-
   Beyond warranty : building your referral business / Carol Smith.
      p. cm.
   Rev. ed. of: Warranty service for home builders. c2002.
   Includes bibliographical references.
   ISBN 978-0-86718-632-1
   1. Construction industry--Customer service. 2. Warranty--Forms. I.
Smith, Carol, 1946- Warranty service for home builders. II. Title.
HD9715.A2S5467 2008
690' .80688--dc22

                                    2007050006

For further information, please contact:

BuilderBooks.com® A Service of **NAHB** NATIONAL ASSOCIATION OF HOME BUILDERS
BOOKS THAT BUILD YOUR BUSINESS
National Association of Home Builders
1201 15th Street, NW
Washington, DC 20005-2800
800-223-2665
Visit us online at www.BuilderBooks.com®

To

William Young
NAHB's director of Consumer Affairs

from
1983 to 2001

colleague, mentor, friend, and
one of the people who "get it"
where customer service is concerned

Contents

List of Illustrations
(An asterisk indicates the illustration appears on the CD.)

Chapter 4 Warranty Service Structure

Chapter 5 Decisions

Chapter 6 Repairs

Chapter 7 Special Handling

Chapter 8 Closure

Chapter 9 Denials

Chapter 10 Trade Contractors

Chapter 11 Warranty Reports

Chapter 12 Warranty Office Routine

Chapter 13 The Service Horizon: Beyond Traditional Warranty

Chapter 14 Staffing for After-Move-In Service

About the Author

Internationally renowned, Carol Smith is a customer relations expert with nearly 35 years of experience in the home building, customer relations, real estate, and mortgage industries with home owners and builders alike. A best-selling author of nearly a dozen books focusing on home building, Carol's successful books include *Buying Your Home: An Insider's Guide*, 2nd ed.; *Customer Service for Home Builders*; *Meetings with Clients: A Self-Study Manual for a Home Builder's Frontline Personnel*; *Homeowner Manual: A Template for Home Builders*, 2nd ed.; *Dear Homeowner: A Book of Customer Service Letters*; and other products.

She founded and publishes *Home Address*, a newsletter for builders. She founded the Association for Customer Relations Professionals and speaks to standing-room-only audiences at the International Builders' Show of the National Association of Home Builders, at state and local builders' association functions, and at meetings in Canada, Australia, and New Zealand.

She has been married for 22 tempestuous, exciting, wonderful years—most of them spent remodeling their home. They are now building what will be their last(?) home.

Acknowledgments

BuilderBooks especially thanks Sam Bradley, Sam Bradley Homes, Springfield, Missouri; Yvonne Rawson, Pueblo Builders, Las Cruces, New Mexico; and Bob Whitten, SMA Consulting, Orlando, Florida; for their strong support for this book in the proposal stage. Special thanks also to David Crump, NAHB Director of Legal Research, for his assistance regarding warranty and maintenance standards and guidelines and the Notice and Right-to-Repair Laws.

The author especially thanks Tracey and Chuck Gunderson of HOMsoft™, Burnsville, Minnesota, for providing the computer screens used to illustrate how software can facilitate the processes described in this book. These screens serve as examples only; their use does not imply endorsement, recommendation, or favored status by the National Association of Home Builders. She also especially thanks Russell Nassof, Environomics/TRC for the use of his five elements for dealing with water intrustion in Chapter 7, "Special Handling."

Thanks to the hundreds of industry professionals who over the years through their questions have challenged me to develop new insights, allowed me to observe their procedures in action, and shared thousands of great ideas—large and small—many of which are reflected in the pages that follow. Because of their continuous efforts to evolve to new levels of excellence, I have been privileged to circulate suggestions back to readers of this book.

Introduction

Much has changed in the years since BuilderBooks first detailed how to organize warranty service for home owners in a book by Carol Smith titled, *Warranty Service for Home Builders,* This book, *Beyond Warranty: Building Your Referral Business,* tells you how to do that and more. New suggestions address other after-move-in services besides warranty. To understand why you need this information, you have to take stock of the service context within which builders now work.

The Satisfaction Challenge

Many home building veterans—including this author—have observed that consumers have become tougher to satisfy, quicker to anger, more sophisticated, and sometimes downright unreasonable in their expectations. While all of these observations are true, they stop short of acknowledging that those customers often had good reason to express frustration.

In large part, today's home buyers have lost patience with inadequate planning, haphazard communication, sloppy follow-through, lack of empathy, unresolved quality issues, incomplete homes, and policies that defy common sense. Further, they are exasperated with written materials that overwhelm with a clutter of information. Related professions have prospered on the resulting disagreements—from private home inspectors, to attorneys, to nearly militant consumer groups who seem to want all builders jailed.

To be fair, a few opportunistic customers enter the new home experience hoping for a chance to take advantage of their builders. However, most home buyers simply want what they were sold, regular communication, and a bit of empathy. In response to these consumers, builders are taking these new customer service trends into account as they plan their operations.

Customer Service Trends

Achieving home owner satisfaction by first aligning and then *meeting* customer expectations was sufficiently challenging, but that target has now changed. The new objective is to go beyond satisfaction by first aligning and then *exceeding* customer expectations.

This goal was initially termed *customer delight*. Savvy builders soon clarified that *delight*, while appealing, falls short of conveying the depth of the changes that were about to take place. In fact, home builders now target genuinely positive long-term relationships with loyal home owners. This objective goes well beyond a few smiles, cookies, or balloons; it requires sincere respect, appreciation, and attention to every detail. Builders needed to define expectations more carefully and transform many traditional builder processes.

Many builders had opportunities to improve the fundamental step of aligning home buyer expectations. When you know something well, remembering what not knowing it was like is difficult. Examining the new home experience from the home buyers' perspective has allowed builders to identify and articulate more precisely the expectations their home buyers need.

At the same time that builders assembled more complete and reliable information about their products, processes, and services, they saw what home buyers meant when they said, "Love the house; hate the process." Delivering a beautiful home also no longer proves to be sufficient; builders must deliver a memorable experience as well. *Managing the experience* serves as the catch phrase for streamlining processes, coordinating events into a cohesive whole, and adding some fun touches.

Customers should enjoy the new-home experience, not merely survive it.

Touch point is shorthand for the opportunity created when anyone in your organization has contact with a customer. The term should remind builders of the potential impact of those contacts—positive or negative. The touch-point concept serves as a tool for identifying, analyzing, and improving events in the process—the main building block of the managed experience. A few builders need to add more interactions with customers, but for most companies the overriding task is to manage their existing touch points more effectively. Carefully planned and well-executed touch points add up to a memorable experience.

Two other powerful forces have converged on the home building industry: customer satisfaction surveys and the Notice and Right-to-Repair legislation. To paraphrase a movie title, these forces create what you could call, "a perfect service storm" for builders.

Home Owner Satisfaction Surveys

Third-party survey companies have polled tens of thousands of home owners and publicized summaries of the results in many cases. These statistics confirm what any experienced warranty manager could have told you: After-move-in service is vital to a company's reputation and future business prosperity. Home owners' opinions of after-move-in service are one of the most significant factors in their decisions to refer potential home buyers to their builders.

Notice and Right-to-Repair Laws

Many states have passed or are in the process of passing *notice and right-to-repair* legislation of some type. The legislators hope to reduce conflicts by defining construction defects and prescribing processes for handling warranty claims. Builders cannot rely solely on legislative action to establish good relations with home buyers: In spite of these efforts, misunderstandings and disputes can persist; healthy customer relationships require hard work and attention to detail by builders.

Increased Planning

You cannot discuss warranty and other after-move-in services in a vacuum. Rather, you must consider them as parts of your overall service system. Mention warranty service and the conversation quickly moves to expectations, home quality, move-in condition, and a host of other details.

Many customer interactions take place before the home owner reaches the warranty period. The warranty office inherits the customer's expectations as well as the quality of the home. The fastest, most-efficient warranty service will fail to please home owners if the sales staff has overpromised or if construction was so poorly executed that the home owner must call upon the warranty office for dozens of repairs.

Hundreds of successful builders have shown that meeting the following goals reduces warranty work, creates happier customers, and produces greater profit:

* accurate customer expectations
* effective quality control
* a commitment to delivering a complete and clean home

You would spend fewer dollars on repairs and on advertising as previous home buyers become part of your sales team. Builders with enviable reputations are able to get top dollar for the quality and service they deliver with their homes. You don't get there accidentally; you need a plan.

An overview of what this plan might look like appears in Chapter 1, "Customer Expectations." It provides an essential backdrop for the after-move-in details that follow.

Warranty Excellence

Minimally, a builder's warranty service strives to meet legal obligations and avoid conflicts. Tremendous benefits accrue when builders expand warranty's role beyond these minimums. For instance, by monitoring the number and nature of warranty items, builders can identify opportunities to improve their product, processes, and staff training. This effort brings higher quality product, happier home owners, and lower repair costs. Procedures that impress home owners are precisely planned and rigorously carried out.

Chapters 2 through 12 offer a detailed blueprint for warranty service, from documents through organization of the department (or desk in the case of small companies). A few years ago, the discussion of warranty service ended at that point. Now you must look beyond warranty to find other opportunities for service that brings referrals.

Beyond Warranty

Keep in mind that the goal now is to exceed customer expectations. This goal prompts the suggestion for more after-move-in services than warranty alone. However, you cannot use the new procedures such as those discussed in Chapter 13, "The Service Horizon: Beyond Warranty," as a substitute for effective warranty service. If warranty service proves to be slow and unprofessional, is peppered with missed appointments, and involves Band-Aid repairs, no amount of additional attention will rescue the relationship. In this context, you can define after-move-in service as warranty service excellence, plus planned additional attention unrelated to warranty matters.

The object is to establish and maintain a healthy, long-term relationship with home owners. The after-move-in services approach is for builders who accept on instinct, logic, and experience that excited home owners help sell homes, increase profits, and make everyone's workday gratifying. The first step to getting referrals is to deserve them. This situation requires that home owners believe they received good value for their housing dollar and that they were treated well. After-move-in services contribute significantly to both the intellectual and the emotional sides of home owner opinions.

Staffing

Staffing for after-move-in service presents some unique issues. These issues are identified and discussed in Chapter 14. However, you ultimately decide to organize these tasks, whomever you trust with the assignments cares for your home owners and your reputation. To succeed, they will need training and technical support.

Training

The best of intentions bring success only when a well-trained staff carries them out with confidence and enthusiasm. Toward that end, this book includes, where applicable, suggestions for training activities to expand your team's insights and skills. The ideas presented here are by no means exhaustive, but they will get you started and stimulate other ideas.

Technology

Warranty work and customer relationship management both lend themselves to data bases, forms, reports, checklists, and mail merges. Although this book does not endlessly repeat, "Information technology can help you do this task," rest assured that it can. The right equipment and computer systems can translate the concepts discussed here into e-media, manipulate them electronically to communicate more quickly, keep records more accurately, and make tracking more productive.

> After-move-in service has become the new service frontier.

The guidance systems offered here provide builders with insights and tools they need to navigate this new service frontier successfully. Use them to explore the territory and create benefits for your company by creating benefits for your home owners.

Using This Book and CD

The CD with *Beyond Warranty: Building Your Referral Business* contains forms and documents from the book that you can easily customize for your business needs. To access the forms, follow these simple steps in order:

* Place the CD in your CD drive.
* Launch Microsoft Windows Explorer.
* Click the icon that represents your CD drive.
* Choose the appropriate file type for your word processing system. The documents are available as both Microsoft Word files and PDF. If you use Microsoft Word as your processor, the Word files will appear exactly as they do in the book.
* Find the document you need in one of the following chapter subdirectories, which match the contents page:

  1. Home Owner Expectations
  2. New Home Warranty
  3. Warranty and Maintenance Standards and Guidelines
  4. Warranty Service Structure
  5. Decisions
  6. Warranty Repairs
  7. Special Handling
  8. Closure
  9. Denials
  10. Trade Contractors
  11. Warranty Reports
  12. Warranty Office Routine

13. The Service Horizon: Beyond Traditional Warranty
14. Staffing for After-Move-In Service

* Locate the file you want to use. The file names correspond to the names in the List of Illustrations. For example, in the subdirectory, "Chapter 1, Home Owner Expectations," you would find files for Figure 1.1, "Home Owner Entry Guide: How Do Your Get Home Owners to Read It," and Figure 1.2, "Meeting Matrix," both of which are figures in Chapter 1.
* Open the document you want to use. If you select the Word file, you will notice that the document opens as a "Read-only" file. Simply select the "Save as . . ." function, listed under the File menu, and save the document to your computer.
* In the documents, right-angle brackets, [Builder], indicate a place to enter information to customize the document for your business. You can then save the document, and each time someone uses that document that information will appear in it. The less-than and more-than brackets, <product name> indicate information you would change each time you use a document to customize it for a particular customer.

Now you are ready to customize the files to suit your business needs.

Please note that the sample warranty documents in this book are provided for educational purposes. They should not be used as forms. Warranty documents are designed to cover the major topics of consideration for such documents. However, they do not and cannot apply to every situation, nor do they comply with any particular state law. Some of the items included will not apply to a particular situation; while in other cases, additional terms may be appropriate. Laws can vary, and some states may require specific language and formats for certain topics. Builders should work with their attorneys to prepare warranty documents that meet their particular needs.

Please also note that throughout the text of this book, the term *warranty representative* has been shortened to *warranty rep* because that is the term that builders and their staff members use daily.

Home Buyer Expectations

Aligning customer expectations is a process that takes place over time and involves all functions of your company. The tools available for accomplishing this vital and complex task include the following:

- a home owner guide
- planned customer meetings and/or home buyer seminars
- status updates, typically weekly
- a series of routine letters that both remind home owners of actions and confirm them
- a system for documenting customer-initiated communications

As builders organize the materials that support this effort and train personnel, they should think in terms of three main topics: product, process, and services. When a company addresses all three aspects of the home buying experience, it has its best chance of success.

Home Owner Guide

To be effective, your home owner guide must be attractive, comprehensive, well-organized, and user-friendly. BuilderBooks, the publishing arm of the National Association of Home Builders, has published a model, *Homeowner Manual: A Template for Home Builders* by Carol Smith, to give you a strong basis for customizing your material. The front section, "How to Use this Model" provides step-by-step suggestions for customizing and using it.

Establishing the authority of the company home owner guide is a team effort and must take place over time. Figure 1.1 offers some suggested uses to get you started. By the time the home buyers reach warranty, they are accustomed to and expect the company personnel to find direction in the home owner guide.

Most of us accept the authority of the dictionary because we have seen others around us accept its authority over and over. Your company can establish the authority of the home owner guide in the same manner by showing home buyers that the company uses the home owner guide and does as it says.

FIGURE

1.1 Home Owner Guide Entry: How Do You Get Customers to Read It?

They won't read it, any more than any of us read a dictionary. Think of the home owner guide as a reference, like the dictionary: a place to find answers. The secret to success with a home owner guide is to establish its authority just as society established the authority of the dictionary in all of our minds—we all saw many other people, over time, used the dictionary and followed what it says. Integrate your guide throughout the process in the same way, let customers see frontline personnel use it and follow it. Suggestions on how to accomplish this task are listed below.

Read. Ask all staff to read it—even the receptionist, accounts payable, and payroll administrator. Everyone. Let them know this is a living document, subject to revision and updating. Therefore, each staff member should create a "Home Owner Guide Revisions" file and make note of their ideas for future updates.

Mark. As they read or work with the guide, company personnel will benefit greatly from marking their copies. They should highlight key points to show buyers and add paper clips or sticky notes to pages so they can locate information they need to share with customers quickly and easily. Staff copies should look used.

Display. Once company personnel are familiar with it, the next step is to help buyers become familiar with it. Begin by displaying your guide at your sales and selection centers; place a copy on the kitchen counter in inventory homes.

Mention. Mention the guide in the sales presentation. This first mention should be casual and brief. "When you buy one of our homes, you'll receive a copy of our home owner guide. You're welcome to look through the display copy if you like."

Deliver. Deliver the buyers' copies at contract. The most effective guides help buyers throughout the process by presenting routine information in a concise and well-organized format. Buyers derive a sense of things being under control when they can see what's coming up next. They are also impressed with the professionalism of the builder.

Review. Review the guide briefly at delivery, pointing out the overall organization and topics covered. This review should take 3 to 5 minutes. This step is the best time to recommend the following points to buyers:

- Read through maintenance information prior to making selections. Buyers understanding of maintenance tasks involved with various features and finish materials may influence their choices.
- Bring the guide to all scheduled meetings.
- Store documents and even color samples in the guide for convenient reference.

Acknowledge. At delivery, ask buyers to acknowledge receipt of the guide. It can be mentioned in the contract or listed on the buyer's contract checklist. Some builders include a clause that allows the buyers to cancel the contract within 72 hours if they object to anything in the documents. Such a clause makes people realize the importance of this material. You will rarely lose a buyer, and if you do, it is probably for the best.

Assign. The salesperson should suggest to buyers that reading the first few sections (finance and selections) will help them navigate the initial steps in the process.

Remind. Ask buyers whether they have had a chance to read their guide and whether they have any questions about it. Let them know prior to each meeting

FIGURE
1.1 *Continued*

(when setting up the appointment) that "details about this meeting and a copy of the agenda we will cover can be found on page <x> in your home owner guide."

Refer. When questions arise, refer to the guide for answers. "I believe that's covered in our guide. Let me look it up for you. Well, here on page 18 it says . . . so that would mean we will . . . ." You are showing buyers by your example that the answers they need are in the guide.

Discuss. Include one of more references to the guide in every routine meeting. At the frame tour, open it to the page describing that meeting and say, "We've come about halfway through this process; here's where we are today, doing your frame tour." At the end of each meeting, open the guide and show buyers where they can find information about the next meeting. At the orientation, demonstrate how easy it is to locate maintenance and warranty information. As the tour of the home progresses, mention several times that information presented is covered in writing in the guide for later reference.

Carry. When buyers see the builder's staff with their home owner guides on their desks, in their vehicles, and under their arms, the buyers realize they use this information. Warranty reps can carry a copy to warranty inspections as a reference (tempered by common sense, of course).

> "I believe this item is discussed in our home owner guide. Let's look it up . . . Yes, right here the book says we will . . . and as a home owner your maintenance on this item includes . . . ." Because the book is doing the talking, the warranty rep does not appear to be making a personal choice.

Quote. In extreme cases, quote the guide in follow-up letters to home owners and include a copy of the page you are referring to with your letter.

Feedback. Asking customers to evaluate the book can serve as another way to stimulate their interest in it and often generates some great ideas for future revisions. One home owner in Florida suggested a chart or table summarizing suggested maintenance routines on one page.

Revise. Annually, collect notes from staff, comments from home owners, and suggestions from the trades. Update the contents to keep information current and accurate. Generally warranty heads up this task as it has the largest section of the book. All departments should update their own sections, and one person should edit new material for style consistency and friendly tone.

File. Label each edition with the date you begin using it. Then file a master for future reference. As future revised editions are completed, repeat this step. Keep another working office copy for adding notes and suggestions for future changes.

Customers learn that the company does what the guide says because that is what happens throughout the building process. When this process is done well, warranty inherits home owners with appropriate expectations. As a result the warranty staff receives fewer questions to which the answer is no.

Techno-savvy people appropriately ask, why not put the home owner guide on a CD? That's fine, but you still need a hard copy to carry under your arm as you tour the buyers' home at frame stage and so on. Home buyers need to see builder personnel use the book many times in a variety of situations. You can help make this happen by beginning with a guide that is attractive, comprehensive, well-organized, and user-friendly.

Attractive

The appearance of your guide, including every detail from the quality of the cover to consistent formatting, helps determine how much attention the binder will receive. Your salespeople must be proud to mention it and deliver it. The salesperson's presentation of the guide and use of it from the beginning is the critical first step in managing customer expectations. Home buyers should believe they are receiving something useful and valuable. Lack of planning, poor-quality paper, or a mishmash of fonts and font treatments diminish the impact of the information the guide contains. Regardless of company size, a polished home owner guide is tangible evidence of an established and well-organized business process. Owners and staff members of small-volume companies will find this tool essential for differentiating their companies from their competition.

Comprehensive

As with a dictionary, this volume is a reference. Customers are unlikely to read the guide from cover to cover (although your personnel should). Your goal is to get customers to turn to it when they want information. Therefore, your home owner guide must contain answers and details, all the information customers need to navigate the new home process successfully. (A dictionary is of little use if every time you look up a word, you find it is not included.) This necessity means describing every step in the process. For most builders this mandate means including a discussion about each of the planned buyer meetings along with supporting materials.

Well-Organized

Again comparing this volume to a dictionary, strong organization is essential. Home buyers and company personnel need to be able to retrieve information conveniently. (Imagine a dictionary that listed definitions in random order rather than alphabetical.) Customers have little patience with informational clutter and will either become frustrated with it or stop using it altogether.

User-Friendly

When this material is well-organized, finding answers is easier. That ease provides a strong start toward making the information user-friendly. Go beyond that beginning to ensure that the tone of the writing is positive and that the details included are accurate and genuinely helpful. Include index tabs printed with the section titles on them. Make the first page of each section a bulleted list highlighting the points covered. This list serves as an essential tool for salespeople in the initial presentation of the binder. Digital photos, cartoons, or a bit of color can make the pages interesting as well.

Planned Customer Meetings

Customer meetings serve two general purposes: They provide predictable opportunities to reinforce information in the guide, and they form a framework for regular communication with home buyers. The Meeting Matrix that appears in Figure 1.2 offers a systematic approach to working with home buyers. You need to consider two dozen meeting fundamentals for every meeting.

Note that the number or names of the meetings vary slightly from builder to builder. Regardless of whether you have a few more, a few less, or whether your meeting names are different, the systematic approach will bring order to those meetings.

When you have responded to each item for each meeting, you will have created a consistent plan along with powerful training and cross-training material. The matrix divides the planning into five main sections: purposes, scheduling, preparation, conducting, and follow-up. Each in turn involves some details that are discussed briefly in the following pages. For detailed discussion of many meeting issues, refer to the book, *Meetings with Clients*, by Carol Smith and published by BuilderBooks.

Purposes

You should describe each meeting in your home owner guide. Ideally, you would format the practical details for each meeting in a uniform manner (see Figure 1.3 for an example). The purpose or purposes of the meeting are listed immediately following the name of the meeting. When the parties participating in the meeting are aligned on the purposes and agenda, they will be more likely to use the time invested efficiently, and the meeting will produce better results: buyers who understand the process.

Schedule

Scheduling questions have no universally right answers. However, all staff members working in the same community should agree on the answers. Such cohesiveness eliminates confusion for company personnel as well as home buyers. (The numbers of the following subheads correspond to the lines on the form in Figure 1.3.)

Triggers (line 1). Typically, either certain paperwork is in hand (for instance, the company has the building permit or the customer has signed all selection sheets before you schedule the preconstruction meeting) or the home has reached a defined stage of construction (rough mechanicals are complete for frame stage tour).

Location (line 2). Some of the items under this subhead are obvious, but again, every staff member working in a community needs to answer this item the same way. In addition, consider the following conditions necessary for effective meetings:

- Choose a meeting room that provides appropriate privacy.
- If possible, forestall interruptions; include reminders to silence cell phones.

FIGURE
1.2   **Meeting Matrix**

Meeting	Contract	Mortgage	Selections 1	Selections 2
Purposes				
Scheduling				
1. Trigger(s)				
2. Location				
3. Attendees				
4. Appointment set up by				
5. Available days and times for appointments				
6. Minimum notice for buyers				
7. Typical duration				
8. Key points to mention in setting up appointment				
9. Details in Home Owner Guide—page				
10. Confirming letter?				
Preparation				
11. Attach appropriate agenda				
12. Review file				
13. Confer with colleagues				
14. Standard items	Agenda Buyer's file Home Owner Guide Business cards	Agenda Buyer's file Business cards	Agenda Buyer's file Home Owner Guide Business cards	Agenda Buyer's file Home Owner Guide Business cards	Home Owner Guide
15. Items particular to this meeting				
Conduct				
16. Personnel introduction				
17. Meeting introduction				
18. Positive comments for opening				
19. Responses to controversial issues				
20. Proactive touches: What memory points have you planned to surprise the home buyers?				
21. Positive comments for conclusions				
22. Conclusion: Key points about next meeting, Home Owner Guide page reference; questions in the meantime				
Follow up				
23. Follow-up letter?				
24. Confirmation of resolved issue(s)				

F I G U R E

1.2 *Continued*

Preconstruction	Frame tour	Orientation	Closing	Warranty visit	Warranty visit
Agenda	Agenda	Agenda	Agenda	Agenda	Agenda
Buyer's file	Buyer's file	Buyer's file	Buyer's file	Buyer's file	Buyer's file
Home Owner Guide	Home Owner Guide	Home Owner Guide	Home Owner Guide	Home Owner Guide	Home Owner Guide
Business cards	Business cards	Business cards	Business cards	Business cards	Business cards

FIGURE

1.3 Home Owner Guide Entry:
Model for Information About Planned Meetings

Meeting	<name of meeting>
When	• <typical time frame> • Between <x> and <y>, Monday through Friday or <q> and <z> Saturday or Sunday
Where	<sales office, new home, design studio, or so on>
Attendees	• Home buyers • <builder staff person> • <builder staff person, if applicable> (Real estate agent is welcome but not required to attend.)
Length	<typical time frame or range>
Purpose	<purpose(s) of the meeting>
Preparation tips	• Read through section <x> of this Home Owner Guide. • Make note of your questions. • Bring this Home Owner Guide with you to the meeting; we will be referring to it and adding new documents. • We will need your undivided attention so plan to attend alone, if possible (without children, other relatives, or friends).

* Choose a space not affected by loud noises and, where possible, close doors to reduce the noise level and increase privacy.
* If necessary, provide temporary supplemental lighting to make reading documents easier.
* For selection appointments and the preconstruction meeting, arrange for adequate table space for reviewing materials.
* Make sure restrooms are available.
* Decide whether to provide refreshments, and if so, where and when.
* If necessary, provide directions and parking instructions to attendees.

Attendees (line 3). The salesperson obviously handles the contract, and the lender addresses the mortgage application, but who manages selections? Will the salesperson participate in the preconstruction meeting? In some companies, construction conducts the frame tour while in others a warranty rep meets with the buyers. The orientation presents similar choices.

Appointment Set Up By (line 4). Ensure that one person handles this step for each meeting to avoid having two or more company representatives contact the buyers. Such missteps send the wrong message to buyers.

Available Days and Times for Appointments (line 5). Each meeting is likely to have different days and times when appointments are available depending on which staff members are involved.

Offer two or more time slots: "We offer preconstruction meeting appointments Tuesday through Thursday at 10 a.m. or at 2 or 4 p.m. Which of those times will work best for you?" Determine the times available by eliminating set obligations such as weekly staff meetings from the calendar; then decide from the days that remain what works best for company personnel. You want to offer as much flexibility as possible to home buyers without causing scheduling conflicts for staff members. You can list the days and times as one of the standard details provided in the home owner guide (see Figure 1.2). Avoid contacting a customer to schedule a meeting and asking, "When would you like to come out?" Too often the buyer will say, "How about 9 a.m. on Saturday?" To which the answer is likely to be, "Wrong, guess again!"

Salespeople can foster customer satisfaction when they thank home buyers for choosing to do business with the company. "We appreciate the confidence you're placing in us. We realize this decision represents a financial and emotional investment, as well as an investment of your time. We plan several meetings with our buyers, some during construction and some after you move in. Many of them are scheduled during normal business hours. We realize we're disrupting your normal schedule so we work hard to make our meetings efficient, interesting, and helpful. I hope you'll take advantage of them."

Minimum Notice (line 6). Naturally the more days ahead you can schedule a meeting the better, but do advise the buyers that the frame stage tour occurs within a small window of time, and notice for that opportunity will be just a day or two.

Typical Duration (line 7). Within a community each kind of meeting will settle into a predictable amount of time. Keep in mind that some customers are naturally more detail oriented (engineers are a classic example) so you need to allow extra time on schedules to avoid having to rush.

Key Points to Mention (line 8) and Home Owner Guide Pages (line 9). Work continuously at creating expectations for the next step. In this case, when you are setting up the appointment for each meeting, mention to the customer that "Page <x> in your Home Owner Guide describes the <meeting being scheduled>. Some important points to keep in mind include . . . ." Your goal is assuring the buyers' comfort level and helping them get the maximum benefit from the time they spend. Additionally, include a copy of each meeting agenda in the home owner guide. Consider adding a watermark to each that says SAMPLE so that as they evolve or change slightly from community to community you do not need to replace them in the binders. Seeing typical agendas shows customers the level of detail you address in these meetings and demonstrates your company's respect for the home buyers' desire to know what is going on.

Confirming Letter (line 10). The routine letters that are part of this system might include some that confirm an appointment and reiterate key preparation details, especially when some tasks need to be done by the buyers and timing permits a letter to be prepared and received via e-mail or the Postal Service. The closing is another excellent example: Buyers are likely to need to confirm last minute financial details, arrange hazard insurance, and transfer utilities.

Prepare

Company representatives need to invest some time and thought into each meeting. The steps are straightforward and, when they are followed in a disciplined manner, they produce outstanding results.

Agenda (line 11). The actual agenda used during the meeting should be printed on no-carbon-required (NCR) paper so that the buyers can have a copy immediately at the end of each meeting. Staff should begin meeting preparation by reviewing the agenda to get topics freshly in mind.

Review File (line 12). If applicable, organize materials in the file in the order in which they will be used at the meeting. If you will give copies of some of them to the buyers, make the copies before the meeting. Double check that the home buyer and the appropriate company official have signed all pages that should be signed and that the home buyer has made any payments due. If any such detail has been overlooked, tag it for attention at the upcoming meeting.

Confer with Colleagues (line 13). Depending on what is appropriate for the meeting scheduled, the superintendent may want to chat with the salesperson; the warranty rep might check in with construction and design. The object is to be aware of any issues that the buyer might mention. Final answers may not be available yet, but you need to be informed about the status of any such items.

Standard Items (line 14) and Items Particular to This Meeting (line 15). Each staff member should designate a storage place for items used at the meetings in which he or she participates. This practice reduces stress and ensures that materials or tools will be on hand. For instance, for the frame tour, the company rep might want to carry a tape measure and have easy access to a level—just in case a question arises. Whether you use a container in a vehicle, a closet shelf, or a desk drawer—establish a specific place to hold these tools so they are readily available when the time comes to go to a meeting.

Conduct

With thorough preparation, conducting customer meetings should be enjoyable. With confidence high and knowing that buyers want or need the information to be covered, staff members can experience gratification from the time invested.

Personnel Introduction (line 16). If two or more staff members are involved in the meeting, the buyers are often meeting one of them for the first time. For example, the salesperson and the superintendent may both participate in the preconstruction meeting. The salesperson should be prepared to introduce the superintendent to the customers with a few sentences that establish credentials.

Meeting Introduction (line 17). Start all meetings the same way. These meetings disrupt the home buyers' normal schedule. Begin every meeting thanking them for their time. Next, review briefly the purpose(s) of the meeting. Not only does this review help in controlling the meeting, it gets all parties focused on the subjects to be covered. After a couple of meetings, the customers will feel assured that the company has a solid plan for the process and their comfort level will increase.

Positive Comments for Opening (line 18). Building a new home requires an investment of time, money, and emotion. It should also be a lot of fun. The buyers are not meeting to have root canals performed; make the meetings fun. "Congratulations folks, your home is complete and ready for you to move in!" might be part of the opening for an orientation.

Responses to Controversial Issues (line 19). Builders can prevent later conflicts by identifying subjects that cause disagreement and buyer disappointment. Establish a reasonable policy or standard for each such item. Rehearse how to present that information to buyers before the issue arises. Doing so will save frontline staff and management time because later (unpleasant) conversations are forestalled. Rehearsing these conversations with staff eliminates many "shot from the hip" comments that can offend customers.

Topics for consideration might include change requests at the frame tour, delivery date updates, or private home inspectors. Being first to raise difficult subjects allows the builder to control the tone of the conversation. The company appears forthright. Over time, the list of subjects will grow—along with the resulting training materials.

Proactive Touches (line 20). Smart builders know that exceeding customer expectations will lead to more repeat and referral sales. Go beyond just managing this process and build some terrific surprises into your system. These surprises need not be expensive or time consuming. They should be thoughtful, creative, and fun. Taking the home buyer's photo on the home site at the end of the preconstruction meeting is one example. Another is to have fabulous refreshments available at selection appointments, a high-quality key ring as part of the closing process, and so on. The matrix disciplines the company to include at least one memory point in each routine meeting.

Positive Comments for Conclusions (line 21). As you wrap up the meeting, review any notes taken for follow-up attention. Ask if the buyers have any questions. Mention you will be happy to check on any details for them, express appreciation for their time, and tell them that you look forward to working with them on the next step. Provide each buyer with a copy of the signed agenda.

Conclusion (line 22). As you used the introduction to prepare the home buyers for the current meeting, use the conclusion of each meeting to prepare them for the next regular meeting. Open the home owner guide to the page that describes that meeting and point out the agenda. Mention key preparation points. End by reminding the home buyers whom to contact if they have any questions before the next meeting.

Follow Up

Follow-Up Letter (line 23). Anticipate that every customer meeting will require some follow-up attention. Begin to address noted items while the home buyers are

Small-volume builders are unlikely to find home buyer seminars helpful during the building process. However, many custom builders have discovered that a seminar that provides an overview of the custom building processes is an effective way to meet potential buyers.

walking to their cars. Later, phone, e-mail, or fax experts for answers. As you track down the answers, share them with the buyers and note the answer and the date you told them on the original agenda. In some instances, the issues involved may warrant a written response to confirm your conversations.

Issue Resolution Confirmation (line 24). When you resolve the last item, confirm with the home buyers that you have answered all questions raised at that meeting and file the agenda in the home buyer's file.

Home Buyer Seminars

The more homes you sell, the more hours staff members will invest in customer meetings. At some point the demands on staff members' time become overwhelming and you must either add more staff or find another approach. That other approach may be home buyer seminars.

Content. Select the topics you want each speaker to cover and organize the material; for most seminars, a chronological organization works well. Present policies and procedures in a positive tone. Where applicable, mention the pages in the home owner guide that you are discussing.

Slides. Establish slide-formatting protocols before you develop your slides so you can apply the format consistently and avoid having to redo them later. Use 18-point type or larger as your smallest font size. Using both upper- and lowercase letters maximizes readability. Each slide should cover just one idea or topic with supporting details. The progression from subject to subject should be logical and clear. Follow either the six-by-six (six bullets, six words each) or the five-by-seven rule (five bullets, seven words each) for most of your slides. Digital photos of current communities make your presentations exciting.

Speakers. Select speakers for home buyer seminars based on their willingness to participate and their knowledge of the subjects to be covered. Not everyone has outstanding public speaking skills, and while people can develop these skills, many people are content without them. Because enthusiasm shows in the final result, the best approach is to use volunteers.

Rehearsals. Initially each speaker would practice alone. Next, get the group together, sit around a table, and go through the entire presentation without stopping. Have someone not involved in speaking track the start and stop times for each speaker. Rehearse until everyone believes the presentation is well-polished. Company personnel not directly involved can play "audience," so your speakers can hold a dress rehearsal. Duplicate all conditions, including the meeting location, handouts, refreshments, and audience evaluation. Feedback from company personnel will help you anticipate questions home buyers might ask.

Invitations to Home Buyers. When you have all the details planned, materials prepared, and speakers rehearsed, introduce the seminar to your home buyers and invite their participation. Begin informally with an in-person invitation in the sales office and follow up with a more formal printed invitation.

Scheduling. Home buyer seminars typically would occur monthly, but sales volume and home buyers' availability will determine the precise timing. Aim to have 10 to 40 people attend. Saturday mornings are popular times, with Thursday evenings a solid second choice. Alternating may be an effective combination. Allow about two hours for the seminar: 90 minutes for the presentations and 30 minutes for questions and answers. Provide refreshments and consider thank you gifts for your participants as well.

Assessing Your Success. Ask for participant feedback with three to five survey questions. Track the impact of the home buyer seminars: Anecdotal evidence indicates that home buyers who attend one of these seminars are markedly more comfortable and cooperative in the home buying process than those who do not attend one. Decide how to reach those who did not attend. You might repeat the invitation for the next program, mail them a summary, or offer to host a traditional one-on-one meeting with them.

Status Updates

In addition to planned meetings, builders have discovered that brief updates help buyers stay comfortable—especially buyers who live out of town. The salesperson typically provides these updates weekly. For many builders, contacting the home buyers immediately following the community team meeting (described in the following pages) works best. The salesperson has the most recent information at that point. You can provide updates by phone, e-mail (digital photos are terrific for this purpose), fax, postcard (with a first class stamp), or a note. When little news is available on the home, provide some detail about the community instead. The time invested is paid back many times over in problems being prevented.

Routine Letters

Repetition helps establish expectations. It must be subtle, so offer information in various ways to avoid making customers feel they are being "beaten over the head" with key points. Planning a series of letters offers builders another way of educating, reminding, and guiding home buyers. Limit the number of letters so they can be administered effectively.

Six to eight letters should be sufficient, but choose the topics carefully. For instance, you might send the suggested letters listed below:

One builder got great results by sending six letters that each contained a different coupon: coffee, ice cream, submarine sandwich, and so on. The last mailing included a DVD about home maintenance and a package of microwave popcorn to encourage viewing it. Compare that to another company that sent more than 20 letters to each of 400 buyers. Unfortunately, customers received the wrong letters at the wrong times, and the person in charge of sending them became stressed and grouchy.

- thank you for buying; outline next steps
- delivery date update following frame tour
- preparation for closing
- thank you for buying
- welcome to warranty
- year-end warranty letter

Begin these update letters with something about the customer or the home; avoid the words *I, we,* or the name of the company as the first word. Your letters are more likely to be read if they capture the reader's interest. *Dear Homeowner* by Carol Smith and published by BuilderBooks includes dozens of sample letters.

Ensure that each letter is friendly in tone, includes valuable information, and contains a bit of emotion: "We look forward to having you join our <community name> community." Or "We are honored that you chose <company name> to build your new home."

Customer Initiated Communications

Even with the home owner guide, planned meetings, routine updates, and standard letters, buyers will still have questions, and they will still want to visit their homes during construction on a random basis. Be prepared to document questions that arise from these unannounced visits. Documentation accomplishes several positive goals:

- Customers are impressed when they see the company being thorough and paying close attention to details.
- The practice encourages customers to put their questions in writing—a habit that can then continue through warranty.
- Follow up and document resolution of all issues can prevent future problems.
- Tracking when the question came in and when the answer went out encourages prompt responses.
- Analyzing these questions and responses over time helps you identify subjects that need to be added to the guide or to agendas for meetings with home buyers.
- In the event of staff turnover, newcomers can provide continuous service and avoid making buyers repeat their questions.

Figures 1.4 and 1.5 offer two methods of documenting random issues. You can provide buyers with the Buyer Inquiry form in Figure 1.4 in the guide, and salespeople should have copies available as well. Construction can use the Site Tour form in Figure 1.5 to document items that arise when buyers appear for an unscheduled visit.

Community Team Meetings

The weekly meeting between the superintendent and the salesperson provides a comprehensive review of *every* customer's status. The procedure is simple—open

FIGURE
1.4 **Buyer Inquiry**

[Logo]

Buyer Inquiry

Purchaser ___*John Buyer*___   Date ___*9/6/–*___

Address ___*3399 West Port Ave*___   Community ___*Bailey's Cove*___

Home phone ___*555-555-1234*___   Home site no. ___*3-27*___

Work phone _____   Plan ___*Trinity Bay*___

Cell phone _____   E-mail _____

Question

___*What would cost be to add fireplace to master —*___

___*standard location and fireplace model (just like the*___

___*model shows)*___

_____

_____

My preference is to receive a response by ✓ phone __ fax __ e-mail __ letter *@ home*

Purchaser ___*John Buyer*___

Response

___*9/7 — Mark says we can do change order for the regular*___

___*selective price if ordered and signed by 9/10/ –*___

___*Change order is ready if you want to proceed.*___

_____

☐ See attached detail or letter.

Office Use: Response Time Tracking		
Received:	Date ___*9/6/–*___	Time ___*10:30 am*___
Responded:	Date ___*9/17/–*___	Time ___*9:15 am*___

9/11 – Buyers have not signed fireplace change order. Mrs. B will be out when they return from vacation mid-October.

By ___*René Sales*___

Date ___*9/7/–*___

FIGURE

1.5   **Site Tour**

[Logo]   Site Tour

Name __Cynthia Buyer__   Date __10/17/–__

Address __339 West Port Ave__   Community __Bailey's Cove__

Home phone __555-555-1234__   Home site no. __3-27__

Work phone __555-702-4444__   Plan __Trinity Bay__

Cell phone __555-702-0050__   E-mail __cbuyer@att.net__

Follow-Up Items

(1) Price out fireplace for master BR — standard location and model —
    Mrs. Buyer realizes extra costs apply due to late change.

                                                        10/19/– Done.

(2) Extend patio per change order dated 10/15 — Home buyer has
    signed and paid for design.

                                                        10/18/– Ordered.

(3) Is roof tile color correct? Check w/supplier to confirm.

                              10/19/– Remington has confirmed color is correct.

Office Use: Response Time Tracking

Received: Date __10/17/–__ Time __5:15 pm__

Responded: Date __10/19/–__ Time __8:05 am__

Purchaser __Cynthia Buyer__   Date __10/17/–__

Purchaser _____   Date _____

Builder __Alex Smith__   Date __10/17/–__

and examine each buyer file in alphabetical order. Share new information ("The Thompson's mortgage has been approved."), and recent contacts with the buyer ("Mrs. Harrington expressed concern about the dying tree on her home site. Do we have any plans for taking the tree down?"), or new change orders ("Mr. Smith wants to go ahead with the fireplace. He'll be out Thursday to sign the change order and give us a check.")

Especially if a customer has had no contact with his or her salesperson or superintendent over the preceding week, plan an update. Depending on the circumstances, the salesperson may make a call, send an e-mail, or fax the buyer. Sometimes contact from the superintendent might be more appropriate. Although such updates need not be time-consuming, they can forestall many a crisis.

Certainly urgent questions should be discussed by the on-site team as they arise. But many details are not urgent and discussing them at one predictable time each week has a calming effect on operations. Staff members feel organized, more in control, and less fragmented. Furthermore, home buyers quickly discover that issues they discuss with one team member are conveyed to the others. This knowledge increases the buyer's confidence in the team and discourages that occasional manipulative customer from attempting to play team members against each other.

Community and Product Updates

Weekly on-site meetings address another frequent difficulty for community personnel and can ultimately prevent conflicts with home buyers: staying up-to-date on changes in procedures, product (method or material), or amenity status. Topics that might be part of this routine review include the condition of inventory homes, model maintenance needs, landscaping, and common area details, as well as jobsite conditions and trade contractor relations.

Warranty Participation

Although warranty staff may not be able to attend weekly, the warranty representative (rep) for the community should sit in on this meeting at least once a month. Creating a predictable and reliable opportunity for sales, construction, and warranty to discuss community issues builds mutual respect, keeps everyone focused, and can prevent minor annoyances from escalating into major conflicts.

The warranty rep should arrive with reports in hand, ready to provide a brief overview of the nature, number, and status of work within the community. The warranty rep should bring recurring items to the superintendent's attention along with any trades who are slow to respond. The rep also can review information buyers need from sales about warranty standards or procedures. Sales and construction each have a chance to ask questions, make suggestions, and pass along compliments (or complaints) they may hear from home owners.

Meeting Documentation

A simple agenda used in each community brings consistency to documenting meetings company wide and from week to week within each individual community.

Not only does this consistency make the meetings efficient, but as personnel change to new communities, they already will be familiar with the agenda and can review past notes to become familiar with details about their new communities.

Teachable Moments

Over the long term, monitor questions that arise. When new topics need attention—especially those that need a "teachable moment" approach—ask yourself the following questions:

* Who is the best person or persons to present the needed information?
* When is the best time to address the subject? Certainly you want to cover it before it becomes an issue with buyers.
* What method should the company use? Possibilities include conversation, demonstration, meeting agenda, letter, home owner guide, and signed document. Again, you can use more than one.

In the past, the home building industry did not talk to buyers about radon, low-flush toilets, or mold. Somewhere out there a new topic lurks—watch for it and incorporate new information to prevent your customers from having to ask (or complain).

Volunteering information early and reinforcing it regularly has proven to be an effective combination for reducing collisions of buyer wishes and builder intentions. Beginning in the sales office on the first visit, show prospects that your company has thought through and planned the new home building and buying process and the accompanying services.

New Home Warranty

Many consumers define warranty as that fine print document that expires three days before the item it covered breaks. Dictionaries define it as a guarantee. Builders know that their warranty is a commitment to respond when a part of the home fails to perform within required standards.

Builders can encounter serious legal problems if they lack understanding of their warranty obligations. This chapter provides a general overview of terminology and requirements. More detail can be found in the BuilderBooks publications, *Warranties for Builders and Remodelers* (by David Jaffe, David Crump, and Felicia Watson) and *Contracts and Liabilities* (by David Jaffe and David Crump). However, because warranty laws vary from state to state, builders should consult their attorneys about local requirements for content and format. Some states may mandate specific language, capitalization, font size, and/or specific location in the warranty.

Implied Warranty

Almost every state provides consumers with some degree of implied warranty protection for new home sales. Utah is the only exception. The theory behind the implied warranty obligation is that a buyer is entitled to believe no builder would knowingly sell a defective home. This assumption can include such concepts as the home being "fit for its intended use" or that it will be "habitable." If the implied warranty statutes of your state and your written warranty conflict, a court may enforce the implied warranty statutes.

An implied warranty disclaimer states that the builder and the purchaser have agreed to waive implied warranties and the disclaimer must name the specific implied warranties. You cannot use a blanket disclaimer. Paragraph 9 of the sample limited warranty in Figure 2.1 shows an example of the wording. Courts do not favor such disclaimers and they interpret them strictly. Texas and Massachusetts do not permit disclaimers. Where permitted, the waiver must be in writing, and be clear and conspicuous (large print, bold type, all capitals). Disclaimers should be written by an attorney familiar with applicable state laws. In some jurisdictions, an implied warranty disclaimer will not guarantee the builder protection against implied warranty obligations.

FIGURE

2.1   Sample One-Year Limited Warranty Agreement

[Builder], hereafter called the "Company," whose office is located at 555 Construction Road, City, State, 55555, extends the following one-year limited warranty to <u>John C. and Cynthia A. Buyer</u> hereafter referred to as "Owner," who has contracted with the Company for purchase of the home located at <u>3399 West Port Avenue</u>, Lot <u>27</u>, Block <u>3</u>, in <u>Bailey</u> County, state of <u>Colorado</u>, for the purchase price of $<u>567,952.90</u> <u>(five hundred sixty-seven thousand eight hundred fifty-two dollars and ninety cents)</u>.

The commencement date of the warranty is <u>April 3, 20–</u>, and extends for a period of one year.

1. Coverage on Home Except Consumer Products

The Company expressly warrants to the original Owner and to subsequent Owner of the home that the home will be free from defects in materials and workmanship resulting from noncompliance with the standards set forth in the Limited Warranty Guidelines in effect on the date of this limited warranty, included in the Company Home Owner Guide and which are part of this warranty.

2. Coverage on Consumer Products

For purposes of this Limited Warranty Agreement, the term "consumer products" means all appliances, equipment and other items that are consumer products for the purposes of the Magnuson-Moss Warranty Act (15 U.S.C., sections 2301-2312) and that are located in the home on the commencement date of the warranty. The Company expressly warrants that all consumer products will, for a period of one year after the commencement date of this warranty, be free from defects resulting from noncompliance with the generally accepted standards in the state in which the home is located, that assure quality of materials and workmanship. Any implied warranties for merchantability, workmanship, or fitness for intended use on any such consumer products shall terminate on the same date as the express warranty stated above. Some states do not allow limitations on how long an implied warranty lasts, so this limitation may not apply to you. The Company hereby assigns to the Owner all rights under manufacturers' warranties covering consumer products. Defects in items covered by manufacturers' warranties are excluded from coverage of this limited warranty, and the Owner should follow the procedures in the manufacturers' warranties if defects appear in those items. This warranty gives you specific legal rights, and you may have other rights that vary from state to state.

3. Company's Obligations

If a covered defect occurs during the one-year warranty period, the Company agrees to repair, replace, or pay the Owner the reasonable cost of repairing or replacing the defective item. The Company's total liability under this warranty is limited to the purchase price of the home stated above. The choice among repair, replacement, or payment is the Company's. Any steps taken by the Company to correct defects shall not act to extend the term of this warranty. All repairs by the

FIGURE
2.1 *Continued*

Company shall be at no charge to the Owner and shall be performed within a reasonable length of time, defined as 30 days from the date on any warranty work order issued by the Company unless other scheduling is arranged with the Owner.

4. Owner's Obligation

Owner must provide normal maintenance and proper care of the home according to this warranty, the warranties of manufacturers of consumer products, and generally accepted standards of the state in which the home is located. The Company must be notified in writing, by the Owner, of the existence of any defect before the Company is responsible for the correction of that defect. Written notice of a defect must be received by the Company prior to the expiration of the warranty on that defect and no action at law or in equity may be brought by the Owner against the Company for failure to remedy or repair any defect about which the Company has not received timely notice in writing. The Owner must provide the Company with access to the home during normal business hours, Monday through Friday, 8:00 a.m. to 5:00 p.m., to inspect the defect reported and, if necessary, to take corrective action.

5. Insurance

In the event the Company repairs or replaces or pays the cost of repairing or replacing any defect covered by this warranty for which the Owner is covered by insurance or a warranty provided by another party, Owner must, upon request of the Company, assign the proceeds of such insurance or other warranty to the Company to the extent of the cost to the Company of such repair or replacement.

6. Consequential or Incidental Damages

Consequential or incidental damages are excluded from this warranty. Some states do not allow the exclusion or limitation of incidental or consequential damages, so the above limitation or exclusion may not apply to you.

7. Other Exclusions

The following additional items are excluded from limited warranty:

a. Defects in any item that was not part of the original home as constructed by the Company.
b. Any defect caused by or worsened by negligence, improper maintenance, lack of maintenance, improper action or inaction, or willful or malicious acts by any party other than the Company, its employees, agents, or trade contractors.
c. Normal wear and tear of the home or consumer products in the home.
d. Loss or damage caused by acts of God, including but not limited to fire, explosion, smoke, water escape, changes that are not reasonably foreseeable in the

Continued

FIGURE

2.1 *Continued*

level of underground water table, glass breakage, windstorm, hail, lightning, falling trees, aircraft, vehicles, flood, and earthquakes.

e. Any defect or damage caused by changes in the grading or drainage patterns or by excessive watering of the ground of the Owner's property or adjacent property by any party other than the Company, its employees, agents, or trade contractors.

f. Any loss or damage that arises while the home is being used primarily for non-residential purposes.

g. Any damage to the extent it is caused or made worse by the failure of anyone other than the Company or its employees, agents, or trade contractors to comply with the requirements of this warranty or the requirements of warranties of manufacturers of appliances, equipment, or fixtures.

h. Any defect or damage that is covered by a manufacturer's warranty that has been assigned to Owner under paragraph 2 of this Limited Warranty.

i. Failure of Owner to take timely action to minimize loss or damage or failure of Owner to give the Company timely notice of the defect.

j. Insect or animal damage.

8. Arbitration of Dispute

> Purchaser's Initials *JCB*   Purchaser's Initials *CAB*

The Owner shall promptly contact the Company's warranty department regarding any disputes involving this Agreement. If discussions between the parties do not resolve such dispute, either party may, upon written notice to the other party, submit such dispute to arbitration. The arbitrator shall proceed under the construction industry rules of the American Arbitration Association. The award of the arbitrator shall be final, conclusive, and binding upon the parties. The expenses of the arbitrators shall be shared equally, but each party shall bear its own fees and costs.

9. Exclusive Warranty

> Purchaser Initials *JCB*   Purchaser Initials *CAB*

The Company and the Owner agree that this limited warranty on the home is in lieu of all warranties of habitability or workmanlike construction, or any other warranties, express or implied, to which owner might be entitled, except as to consumer products. No employee, trade contractor, or agent of the Company has the authority to change the terms of this one-year limited warranty.

Dated the _____third_____ day of _____April_____ , __20–__

Owner __*John C. Buyer*_____ Builder __*Bill Builder*_____

Owner __*Cynthia A. Buyer*_____

Express Warranty

An express warranty is made by the builder, either in conversation or in writing. A well-written warranty, supported with clear guidelines, can protect both the buyer and the builder by guaranteeing a certain level of performance for a specified period of time.

Full Warranty

Warranties are either full or limited. Providing a full warranty is almost unheard of among builders. A full warranty must meet the following four tests:

1. Remedies for covered defects are provided within a reasonable time and at no charge.
2. Implied warranty protections are neither waived nor limited.
3. If a defect cannot be repaired after a reasonable number of attempts, the purchaser has the option of returning the home for a refund.
4. The warranty applies to subsequent owners.

A full warranty must be designated as such in the title with the time period mentioned, for instance: "Full One-Year Warranty."

Limited Warranty

Any warranty that does not meet all four full-warranty criteria is automatically a limited warranty. Limited means exactly that: the coverage is limited in some way. The limitations must be clearly described in the warranty and the word *limited* conspicuously printed in the title, for example: "Limited One-Year Warranty."

Magnuson-Moss Act

The Magnuson-Moss Act is a federal law passed in 1975 to provide consumers with certain warranty disclosures when they purchase consumer products. For purposes of this act homes are not considered to be consumer products. However, because homes contain consumer products (dishwasher, furnace, and so on), the Magnuson-Moss Act is significant to builders.

Builders have the option of avoiding Magnuson-Moss requirements by not providing a written warranty, but this situation leaves the builder and the buyers without a clear definition of responsibilities. Another option is to exclude consumer products in the home from coverage. This approach requires the builder to define and list consumer products—a task no one has accomplished with certainty.

Because the Magnuson-Moss requirements are not hard to meet and the options are undesirable, compliance seems logical. Even if you decide to exclude consumer products from your warranty coverage, including the points required by the Magnuson-Moss Act creates a clear warranty. Where applicable, the numbers in parentheses (after the headings listed below) refer to the paragraphs of the sample limited warranty in Figure 2.1. Considerable information is covered in the opening lines of the document, referred to in parentheses as "opening." The sample limited warranty shows one way to respond to Magnuson-Moss requirements. Keep in mind that your state may impose other requirements.

Simple Language and Conspicuous Notice

The Federal Trade Commission's interpretation of the Magnuson-Moss Act requires that warranty information be conveyed in "simple and readily understood" language. The act requires that warranties covering consumer products include specific pieces of information. It also often requires that they be printed in specific words and type sizes.

To Whom Given (opening)

Show the names of the individuals to whom the warranty is given and a clear statement as to whether it is transferable to subsequent owners. The Magnuson-Moss Act does not specify that the warranty must be transferable, but it requires that the warranty make the company's position on this issue clear.

Coverage and Exclusions (1, 2, 7)

The limited warranty should clearly state what it covers and what it does not.

Builder's Obligation (3)

This clause expresses what the builder will do, generally, if a defect is discovered. Builders usually commit "to repair, replace, or pay the reasonable cost of repair or replacement for any covered defect." The choice is solely that of the builder. The builder's obligation is usually limited to the purchase price of the home; so many builder warranties include the price of the home.

Term of Coverage (opening)

This section defines the length of the warranty and when it starts. The materials and workmanship warranty is generally for one year. Some builders include two-year coverage on mechanical systems. The builders modeled these warranties after insured warranty programs. Structural coverage can be included as well. Refer to Figure 2.2 for sample wording that describes structural coverage. Nothing in Magnuson-Moss dictates how long coverage should be; rather, it requires that the time frame be clear to the purchasers. Note that your state may have requirements regarding the type and duration of coverage.

Claims Procedures (4)

For most builders, claims procedures begin with the purchaser notifying the builder, in writing, of defects. Home owners generally can report emergencies by phone. Give the purchaser a grace period—30 days is typical—following the expi-

FIGURE
2.2 Sample Major Structural Coverage

For a period of 10 years from the commencement date of the warranty, the Company expressly warrants to the Owner and any subsequent owner of the home, that the home will be free from major structural defects. A major structural defect is defined as being an actual defect in load-bearing portions of the house that seriously impairs their load-bearing function to the extent that the house is unsafe, unsanitary, or unlivable. For purposes of this definition, the following items comprise the structure of the house:

a. Foundation system
b. Load-bearing stud walls
c. Floor joists
d. Beams, columns, trusses, and rafters

ration of the warranty. Although not a Magnuson-Moss requirement, builder warranties often require the purchaser to maintain the home and allow the builder access during normal business hours for inspections and repairs.

Consequential Damages (6)

Consequential damages are damages that do not flow directly from the breach of contract or warranty; rather, they are a type of indirect damage. If a plumbing leak damages drywall and the dining room table, the damage to the table may constitute consequential damages and would not be covered by the builder's limited warranty. The damage to the drywall would likely be considered direct damage and would be covered by the warranty. Builders can always make exceptions if the circumstances merit doing so.

The Magnuson-Moss Act does not dictate what the builder's position should be on consequential damage, but it specifies that the warranty must clearly state that position for the consumer. Some states do not allow an exclusion of such damages. To call the consumer's attention to this possibility, Magnuson-Moss requires a notice stating "Some states do not allow the exclusion or limitation of incidental or consequential damages, so the above limitation or exclusion may not apply to you." Each builder should investigate this issue with a qualified attorney.

Relationship to Implied Warranties (2, 7)

Although Magnuson-Moss prohibits a complete disclaimer of implied warranty on consumer products, a builder warranty can limit implied warranty coverage on consumer products to the same time frame as the coverage on the entire home. Again, state laws may affect this situation, and the warranty must include a notice saying so.

Dispute Settlement (8)

A clause that establishes a dispute settlement procedure is not required by the Magnuson-Moss Act, but if you include one, the act imposes certain conditions. For example, the decision makers cannot be involved in the manufacture, sale, or service of new homes. A dispute settlement clause can forestall a lawsuit. Generally, the purchaser cannot take legal action until the dispute mechanism has run its course. Have the purchasers initial this paragraph to confirm that they have read it and understand it to avoid later arguments about how to settle arguments.

Customers have begun using the volume of paperwork that builders provide as a justification for litigation even when alternative dispute resolution has been agreed to (and courts seem willing to accept this approach). Several court cases that otherwise would have been referred to arbitration or dismissed entirely have been permitted to move forward by judges because the customer claimed that the builder gave them so much fine print they could not reasonably be expected to read and understand all of it. Home buyers have applied this approach in particular to the waiver of implied warranty and to the alternative dispute resolution clauses. Some builders are addressing this concern by asking home buyers to initial these two clauses to forestall claims that the buyers were unaware of them.

Because the Magnuson-Moss Act itself requires that these notices be "conspicuous," you can put to rest your legal concerns about having customers initial some clauses but not all of them. Initialing simply makes the conspicuous clauses even more conspicuous.

Other Rights (2)

You must include in your warranty the exact words, "This warranty gives you specific legal rights, and you may also have other rights that vary from state to state."

Common Builder Warranty Clauses

In addition to Magnuson-Moss requirements, many builder limited warranties also include the following items:

Builder Name and Address (opening)

This address is where the home owners would mail their written notice of warranty items.

Description of the Property (opening)

Usually this description includes the home's street address, county, and sometimes the lot number, block number, and subdivision.

Consumer Product Warranties Assigned (2)

Coverage for consumer products needs to be assigned to the home buyer. In this clause, reference is often made to consumer products being those items considered to be consumer products for purposes of the Magnuson-Moss Act.

Insurance (5)

This section provides that if the builder repairs something for which the purchaser has home owner insurance coverage, upon request by the builder, the home owner must reimburse the builder for the repair from the insurance payment.

Signatures and Date

The date is the date of closing, unless a rental agreement applies.

Insured Warranty Programs

Insurance-backed warranties offer the protection of having a third party stand behind the home. Builders apply for this coverage and, following acceptance, pay a premium as part of the cost of the home. The policy is issued in the name of the home buyer and commences on the date of closing. If your company provides insured warranty coverage for home buyers, read and understand the terms. Insured warranty programs vary somewhat, but in general, they offer one of the following three types of coverage.

One-Year Materials and Workmanship

The home owner reports items in writing to the builder. The builder screens and repairs the items; any costs are the builder's (or appropriate trade contractor's) responsibility. If a builder is unable or unwilling to provide a repair specified in the insurance policy, the home owner has the insurance program to fall back on. The home owner pays a fee to the insurance company to initiate a claim.

Two-Year Systems

Second-year coverage may be provided for mechanical systems, excluding fixtures such as chandeliers and sinks. The home owner reports warranty items to the builder. The same claims procedure applies if the builder is unwilling or unable to perform repairs specified in the policy.

Structural (usually 10 years)

Insured warranty coverage for structural damage differs from the one- and two-year coverage. The insurance company bears the financial risk of the cost of structural repairs. With some policies the builder's protection begins immediately upon closing. With others, it begins on an anniversary of the closing. In either case, coverage begins immediately for the home owner. What constitutes structural damage for purposes of this coverage is defined in the warranty insurance document.

Reference in Contract

The sales agreement should reference the limited warranty. Figure 2.3 offers an example of the wording for this clause. Because the warranty describes conditions to which the buyers are agreeing, they should receive a copy of the limited warranty and warranty guidelines when they sign the purchase agreement.

Training

Training warranty personnel is an ongoing activity. Annually assess each employee's training needs (with the employee's input), and budget for and schedule appropri-

> ### FIGURE
> ### 2.3   The Limited Warranty: Sample Sales Agreement Clause
>
> **Limited Warranty.** Seller disclaims and purchasers waive the implied warranty of habitability and the implied warranty of workmanlike construction that the residence will be free of latent defects and will be fit for its intended purpose as a home. Seller and Purchasers agree to execute at closing __[name of__ __Company]__ Limited Warranty Agreement. After closing, all claims, rights and remedies of Seller and Purchasers arising out of this contract and Seller's construction and sale of the residence and any consumer products in the residence shall be limited to those set forth in the Limited Warranty Agreement, which is incorporated herein by reference. Such warranty agreement is in lieu of all other warranties of habitability or workmanlike construction and any other express or implied warranties to which purchasers might be entitled. This agreement shall survive delivery of the deed referred to in paragraph __[insert number of__ __paragraph]__ hereof. Purchasers acknowledge that they have received copies of a specimen of the __[name of Company]__ Limited Warranty Agreement and Limited Warranty Guidelines.

ate activities. With regard to the warranty document itself, have personnel read and discuss it twice a year to keep the terms and conditions clearly in mind.

The 20-question warranty quiz in Figure 2.4 may be useful in your training efforts, both with warranty personnel and with salespeople. One goal of this training is to help all parties understand that the complexity of this topic requires accuracy.

Home owner satisfaction studies call attention to the importance of ensuring that buyers understand the provisions of their new home warranties. This effort begins in the sales office with concise and accurate information. Conversations with salespeople about this topic often draw observations such as "Buyers don't ask about warranty" or "Warranty's a negative, I don't like to talk to buyers about that." Quite the contrary, buyers want to know their builder can be trusted to tell the truth, including descriptions of the home buyers' responsibilities. Role play with salespeople to confirm that they are comfortable delivering an enthusiastic and correct description. Rather than seeing warranty as a negative, see it as a matter-of-fact aspect of the builder/buyer relationship.

FIGURE
2.4 **Warranty Quiz**

1. Describe the types and duration of coverages provided by your limited warranty
   a. Materials and workmanship
   b. Systems
   c. Structural
2. What coverage does your state require?
3. Is the warranty insured by a third party or is it self-insured (backed by the Company)?
   a. If insured, who is the underwriter?
   b. What is the underwriter's current financial status?
   c. What fee does the home owner pay to file a claim?
   d. What is the Company's cost for this coverage?
4. If self-insured, how do you respond to prospects who ask why you do not have insured warranty coverage?
5. What is/are the differences between or among the following:
   a. Full warranty and limited warranty?
   b. Implied warranty and express warranty?
   c. Materials and workmanship coverage, systems coverage, and structural coverage?
6. Is the limited warranty assignable (transferable) to subsequent owners?
7. What home owner obligations does the limited warranty impose? What choices does the Company have?
8. What is the limited warranty position on consumer products?
9. Give examples of exclusions listed in the limited warranty.
10. Does the limited warranty include an alternative dispute resolution clause? If so, is it binding?
11. Does the limited warranty contain a waiver of implied warranty? If so, is it enforceable in your state?
12. Does the warranty document contain the price of the home, date of closing, and names and addresses of all parties?
13. Do the home buyers sign the limited warranty at closing?
14. How does your limited warranty compare to the warranties of your competitors?
15. How and when should homeowners report
    a. Emergency items?
    b. Nonemergency items?
16. What is the typical response time for your warranty staff to
    a. Acknowledge receipt of a warranty request?
    b. Inspect items reported?
    c. Accomplish needed work?
17. Which trades provide the best all-around warranty service? The worst?
18. What do you tell customers about your limited warranty?
19. Are oral statements you make regarding warranty coverage binding on the Company?
20. What is the most frequently heard complaint about warranty service? The most frequently heard compliment?

Warranty and Maintenance Standards and Guidelines

CHAPTER

3

Nﾠew home quality standards are a moving target, a tug of war between quality and price, scheduling and skills. A standard can be as clear and concise as a measurement on a ruler or as subjective as a home owner's emotional opinion of a carpet seam. Builders work to assemble a set of standards and guidelines that will appeal to their target market and then strive to produce a product that consistently meets those standards and guidelines at a price buyers are willing to pay.

Sources of Standards and Guidelines

The standards and guidelines builders strive to satisfy in new home construction come from a variety of sources. Most of them exist outside the builder's direct control, such as building codes (which guarantee safety, not quality or aesthetics) and other governmental mandates. Add to these requirements the following items:

* manufacturers' specifications (They assure the proper operation and functioning of the manufacturers' components.)
* industry or professional association recommendations (such as *Residential Construction Performance Guidelines*, third edition, published by BuilderBooks, National Association of Home Builders, and local guidelines produced by local home builders associations)
* environmental and geological conditions such as those related to soil, seismic conditions, weather, termites, and similar factors
* locally accepted construction standards and principles
* trade contractors' standards (the building codes and the specifications of the contract)
* the natural properties of the materials (for example, concrete tends to crack)

Builders follow a significant number of standards and guidelines, and they need to make these standards and guidelines clear to home buyers. You need to write your standards and guidelines (your warranty) in plain English that your home owners can understand. You provide a warranty to avoid trouble and make customers happy. It gives both you and the home owner peace of mind because it protects against construction defects and inadequate materials, spells out the parties' responsibilities, and specifies a method of resolving disputes.

These standards and guidelines fluctuate. They are not static. Codes change; builders, organizations, and jurisdictions adopt and approve new methods and materials; tools and techniques improve; and the market's taste in housing styles, floor plans, and features evolves over time. For the aspects of construction for which builders do have choices, you have an opportunity to identify details that appeal to buyers in your target market. This attention to details provides one more way to differentiate your company from your competitors.

In spite of all the variables and changing factors, builders follow a remarkably similar set of warranty standards and guidelines nationwide. Differences from one company to the next are relatively minor. For instance, one builder caulks concrete flatwork separations when they reach $1/4$ inch, another builder provides this same warranty repair when the separation reaches $3/16$ inch.

With a clear vision of your quality standards, the process of aligning everyone's expectations becomes easier. For you to serve everyone effectively, all participants in the purchase of a home must understand and agree to these standards, guidelines, codes, and law as follows:

* Trade contractors expect to meet the standards.
* On-site supervisors expect to enforce them.
* Salespeople and selection staff expect to present and explain them.
* Home buyers expect to live with them.
* Warranty staff members expect to keep the product within these standards and guidelines.

This chapter primarily focuses on the methods for defining the scope and limitations of warranty coverage for home buyers. The home owner bears responsibility for items that fall outside of warranty coverage as part of normal home maintenance. Therefore, combining information about these topics makes both sets of responsibilities clearer to home buyers. The fundamental tools for this assignment include written warranty coverage and home maintenance guidelines.

Responsibilities Defined

Written warranty standards, increasingly referred to as guidelines in the materials that home buyers receive, define what constitutes a defect and the method a builder will use to correct the item. The builder's limited warranty can refer to these guidelines in wording similar to that shown in paragraph one of the sample limited warranty in Figure 2.1. Maintenance guidelines define common tasks that home owners should perform in caring for their homes.

Common Items

Neither warranty standards and guidelines nor maintenance guidelines are likely to cover every possible event, but they can cover the more common concerns of home owners. Builders can include a statement to this effect in the introduction to their warranty and maintenance guidelines. Figure 3.1 offers an example.

Manufacturers' Use and Care Literature

Builders need to make clear to home buyers that, in spite of sincere efforts to keep their guidelines current and accurate, manufacturers' literature may describe different or additional maintenance steps. Point out this fact in this section of the guide with wording similar to that shown in Figure 3.2.

Organizing Warranty and Maintenance Guidelines

When you organize these guidelines with the end user in mind and present them effectively (and repetitively), they begin to have the desired impact. Written warranty or performance standards and guidelines are readily available from a variety of sources. While well-intended, many of the warranty guidelines in circulation fail to deliver the maximum potential benefits to the builder or the home owner.

This failure results primarily from the physical presentation of the information and how builders use the information. Suggestions for solving the second obstacle appear later under "Gaining Home Buyer Acceptance." The following sections explain common mistakes in the physical presentation of the warranty and maintenance information.

Avoiding Presentation Pitfalls

Warranty guidelines are sometimes written from the perspective of industry insiders and presented in typical construction sequences

Topic Categories and Order. The overall organization of topics may make sense to those in the industry, but to home owners unfamiliar with some of the

FIGURE

3.1   Home Owner Guide Entry: Items Omitted from Guidelines

We recognize that it is impossible to anticipate and describe every action needed for good home care. We focused on items that home owners commonly ask about. The subjects are listed in alphabetical order to make finding answers to your questions convenient. Because we offer home buyers a variety of floor plans and optional features, this guide may discuss components that your home does not include.

In addition to the information contained in the limited warranty itself, this guide includes details about the criteria we will use to evaluate concerns you report. The purpose of this guide is to inform you that our warranty commitment is for the typical concerns that can arise in a new home. The guide describes the corrective action we will take in many common situations.

FIGURE
3.2   Home Owner Guide Entry: Manufacturers' Literature

Please take time to read the literature (warranties and use and care guides) provided by the manufacturers of consumer products and other items in your home. The information contained in that material is not repeated here in this guide. Although much of the information may be familiar to you, some points may differ significantly from homes you have owned or lived in in the past.

We make every effort to keep the information in this manual current. However, if any detail in our discussion conflicts with the manufacturer's recommendations, you should follow the manufacturer's recommendations.

Activate specific manufacturer's warranties by completing and mailing any registration cards included with their materials. In some cases, manufacturers' warranties may extend beyond the first year, and you need to know about such coverages and to include your appliances, equipment, and other products in the manufacturers' systems for any recalls that might occur.

terminology, finding the heading that contains the information they need can be frustrating.

For example, in a widely copied organizational pattern, the home owner who has an issue with cabinets first needs to determine that they are discussed under "Equipment." Similarly, when the grading of their yards fails to drain, home owners are likely to think of the heading, "Grading and Drainage" rather than "Site Work." Wanting information about their fireplaces, the home owners will think of and look for "Fireplace," as a subhead rather than "Miscellaneous" or "Specialties." Needing help with a stucco question, home owners are more likely to think of the subhead, "Stucco," than "Special Coating." Likewise, if a roof concern arises, they will think "Roof" before they think of "Thermal and Moisture Protection."

Deficiency. Columnar formats that include the word *deficiency* at the beginning of each item emphasize that something is wrong with the home—not what the builder needs to call attention to. In addition, the logic behind the order of the descriptions that follow the word *deficiency* eludes the average home owners, who must read through all of them until they come to the one they need.

Multiple Discussions of Unaligned Topics. Builders may present warranty guidelines and maintenance guidelines in two completely separate sections in their home owner guides. Some formats include some maintenance responsibilities mixed in with the warranty guidelines. A second section supposedly lists maintenance tasks, perhaps on home components for which no warranty information was provided in the first section. Yet points about warranty coverage can typically be found mixed into these maintenance descriptions. When the two separate sections fail to discuss the same components, or they refer to the items by different headings and subheads, home owners become frustrated.

Add manufacturers' brochures and booklets, and it is no wonder home owners' heads are spinning. This stew of information makes finding answers to specific questions time consuming because home owners have to look in multiple places and to find answers. Not surprisingly, many home owners give up and list all items they have noticed without bothering to learn who is responsible for correcting them.

Presentation Success

To cure these ills, train yourself to think like the end user. Compose your written materials so that home owners and warranty staff can access information quickly and conveniently.

Complete and Well-Organized. Begin with an extensive list of home components. Organize them in alphabetical order—an organizational approach that is familiar to all home owners and company personnel. Figure 3.3 provides an example of a list excerpted from the contents page of *Homeowner Manual: A Template for Home Builders,* a book written by Carol Smith and published by BuilderBooks.

FIGURE 3.3   Contents Page for Maintenance and Warranty Guidelines

Air-Conditioning	Fencing	Property Boundaries
Alarm System	Fireplace	Railings
Appliances	Foundation	Resilient Flooring
Asphalt	Garage Overhead Door	Roof
Attic Access	Gas Shut-Offs	Rough Carpentry
Brass Fixtures	Ghosting	Septic System
Brick	Grading and Drainage	Shower Doors or Tub
Cabinets	Gutters and Downspouts	Enclosures
Carpet	Hardware	Siding
Caulking	Hardwood Floors	Smoke Detectors
Ceramic Tile	Heating System:	Stairs
Concrete Flatwork	Gas Forced Air	Stucco
Condensation	Heating System:	Sump Pump
Countertops	Heat Pump	Swimming Pools
Crawl Space	Humidifier	Termites
Dampproofing	Insulation	Ventilation
Decks	Landscaping	Water Heater: Electric
Doors and Locks	Mildew	Water Heater: Gas
Drywall	Mirrors	Windows, Screens, and
Easements	Paint and Stain	Sliding Glass Doors
Electrical Systems	Pests and Wildlife	Wood Trim
Evaporative Cooler	Phone Jacks	
Expansion and Contraction	Plumbing	

When builders integrate warranty and maintenance guidelines and present them in a clear order that is easy to consult, they also clarify the delineation of responsibilities between the home owners and the builder.

Maintenance and Warranty Guidelines Together. The discussion of each component should include home maintenance guidelines followed immediately by the builder's detailed warranty commitment (Figure 3.4). This approach makes finding answers to questions easier for the primary end users of this literature: home owners and warranty reps.

Note that the subheads in both the maintenance and warranty guidelines also follow alphabetical order and thus increase their ease of use. Terms such as *deficiency* or *defect* are unnecessary.

Legal or Insurance Requirement?

If a legal or an insurance mandate requires your company to present to the home buyer a document containing only your warranty guidelines, please do provide it. Incorporate the same criteria into the alphabetical version of your warranty guidelines. Explain to home buyers that the first version satisfies a requirement, that the second is formatted for ease of use, and that the criteria are the same.

Organizing Components

Trial and error has spawned several recommendations for maximizing the impact of this material. The following suggestions make the information easier for home owners and company personnel to use.

Format

Establish formatting protocols and use them consistently. For instance, treat safety reminders the same way throughout the document. Use the same size indentations, the same style bullets, and so on. (Builders often box these items.)

Use reasonable margins and incorporate a header or footer; keep paragraphs short. Use spacing, indentations, bullets, and so on to help the eye perceive organization and locate key points as needed. Keep in mind that home buyers are less likely to read a page of solid print in small type and more likely to read a more interesting looking page.

Checklists

Practical tips and genuine short cuts make this material valuable to home owners. Think in terms of helping home owners rather than merely setting limits. For instance, if you sell to retirees who spend part of the year in a second home, they would probably find a list of reminders for extended absences useful (Figure 3.5).

Note the blank lines where home owners can record unique items. This provision protects you from liability if you omitted something relevant to the home owner's lifestyle. Each home owner has unique concerns when leaving his or her home for an extended period; no builder could anticipate all of them.

FIGURE
3.4 Home Owner Guide Entry: Air-Conditioning

Home Owner Use and Maintenance Guidelines

Air-conditioning can greatly enhance the comfort of your home, but if you use it improperly or inefficiently, wasted energy and frustration will result. These hints and suggestions are provided to help you maximize your air-conditioning system.

Your air-conditioning system is a whole-house system. The air-conditioner unit is the mechanism that produces cooler air. The air-conditioning system involves everything inside your home including, for example, drapes, blinds, and windows.

Your home air-conditioning is a closed system, which means that the interior air is continually recycled and cooled until the desired air temperature is reached. Warm outside air disrupts the system and makes cooling impossible. Therefore, you should keep all windows closed. The heat from the sun shining through windows with open drapes is intense enough to overcome the cooling effect of an air-conditioning system. For best results, close the drapes on these windows.

Time affects your expectations of an air-conditioning unit. Unlike a light bulb, which reacts instantly when you turn on a switch, the air-conditioning unit only begins a process when you set the thermostat. For example, if you come home at 6 p.m. when the temperature has reached 90 degrees Fahrenheit and set your thermostat to 75 degrees, the air-conditioning unit will begin cooling, but will take much longer to reach the desired temperature. During the whole day, the sun has been heating not only the air in the house, but the walls, the carpet, and the furniture. At 6 p.m. the air-conditioning unit starts cooling the air, but the walls, carpet, and furniture release heat and nullify this cooling. By the time the air-conditioning unit has cooled the walls, carpet, and furniture, you may well have lost patience.

If evening cooling is your primary goal, set the thermostat at a moderate temperature in the morning while the house is cooler and allow the system to maintain the cooler temperature. You can then slightly lower the temperature setting when you arrive home, with better results. Once the air conditioner is operating, setting the thermostat at 60 degrees will *not* cool the home any faster, and it can result in the unit freezing up and not performing at all. Extended use under these conditions can damage the unit.

Adjust Vents

Maximize air flow to occupied parts of your home by adjusting the vents. Likewise, when the seasons change, readjust them for comfortable heating.

Compressor Level

Maintain the air-conditioning compressor in a level position to prevent inefficient operation and damage to the equipment.
    See also entry for Grading and Drainage.

Humidifier

If a humidifier is installed on the furnace system, turn it off when you use the air-conditioning; otherwise, the additional moisture can cause a freeze-up of the cooling system.

Continued

FIGURE

3.4 *Continued*

Manufacturer's Instructions

The manufacturer's manual specifies maintenance for the condenser. Review and follow these points carefully. Because the air-conditioning system is combined with the heating system, also follow the maintenance instructions for your furnace as part of maintaining your air-conditioning system.

Temperature Variations

Temperatures may vary from room to room by several degrees. This difference results from such variables as floor plan, orientation of the home on the lot, type and use of window coverings, and traffic through the home.

Troubleshooting Tips: No Air-Conditioning

Before calling for service, check to confirm the following situations:

- Thermostat is set to cool, and the temperature is set below room temperature.
- Blower panel cover is set correctly for the furnace blower (fan) to operate. Similar to the way a clothes dryer door operates, this panel pushes in a button that lets the fan motor know it is safe to come on. If that button is not pushed in, the fan will not operate.
- Air-conditioner and furnace circuit breakers on the main electrical panel are on. (Remember if a breaker trips you must turn it from the tripped position to the off position before you can turn it back on.)
- The 220 volt switch on the outside wall near the air conditioner is on.
- Switch on the side of the furnace is on.
- The fuse in furnace is good. (See manufacturer literature for size and location.)
- A clean filter allows adequate air flow. Vents in individual rooms are open.
- Air returns are unobstructed.
- The air conditioner has not frozen from overuse.
- Even if the troubleshooting tips do not identify a solution, the information you gather will be useful to the service provider you call.

[Builder] Limited Warranty Guidelines

The air-conditioning system should maintain a temperature of 78 degrees or a differential of 18 degrees from the outside temperature, measured in the center of each room at a height of five feet above the floor. Lower temperature settings are often possible, but neither the manufacturer nor [Builder] guarantees them.

Compressor

The air-conditioning compressor must be in a level position to operate correctly. If it settles during the warranty period, [Builder] will correct this situation.

Coolant

The outside temperature must be 70 degrees Fahrenheit or higher for the contractor to add coolant to the system. If your home was completed during winter months,

FIGURE

3.4 *Continued*

this charging of the system is unlikely to be complete, and [Builder] will need to charge it in the spring. Although we check and document this situation at orientation, we welcome your call to remind us in the spring.

Nonemergency

Lack of air-conditioning service is not an emergency. Air-conditioning contractors in our region respond to air-conditioning service requests during normal business hours and in the order they receive them.

FIGURE

3.5 Home Owner Guide Entry: Sample Checklist for Extended Absences

Extended Absences

Whether for a vacation, business travel, or other reasons, nearly all of us occasionally leave our homes for days or weeks at a time. With some preparation, such absences can be managed uneventfully. Keep these guidelines in mind and add additional reminders that are appropriate to your situation.

Plan in Advance

- Ask a neighbor to keep an eye on the property. If possible, provide them with a way to reach you while you are away.
- If you will be gone an especially long time (more than two weeks) consider a house sitter.
- Arrange for someone to mow the lawn or shovel snow.
- Notify local security personnel or police of the dates you will be away.
- Stop mail, newspapers, and other deliveries.
- Use lighting timers (available at hardware stores for $10 to $20).
- Confirm that all insurance policies that cover your property and belongings are current and provide sufficient coverage.
- Mark valuable items with identifying information. Consider whether you have irreplaceable items that should be stored in a bank vault or security box.

As You Leave

- Forward phone calls to a relative or a close friend.
- Unplug computers and other electronic devices that might be harmed in an electrical storm.
- Leave window coverings in their most typical positions.
- Confirm that all doors and windows are locked and the deadbolts are engaged.
- Shut off the main water supply. Set the thermostat on the water heater to "vacation" to save energy.
- Store items such as your lawn mower, bicycles, or ladders in the garage.
- Disengage the garage door opener (pull on the rope that hangs from the mechanism). Use the manufacturer's lock to bolt the overhead door. *Caution:* Attempting

Continued

FIGURE
3.5 **Continued**

to operate the garage door opener when the manufacturer's lock is bolted will burn out the motor of your opener. Upon your return, unlock the garage door first, then re-engage the motor (simply push the button to operate the opener and it will reconnect) to restore normal operation.

- Leave a second car in the driveway.
- In summer, turn your air-conditioner fan to on. Set the thermostat on 78.
- In winter, set the thermostat to a minimum of 55 degrees. Leave doors on cabinets that contain plumbing lines open. Leave room doors open as well. The open doors allow heat to circulate.
- Arm your security system, if applicable.

Your Additional Reminders and Notes

_____

_____

_____

Other potential checklists include tips for conserving water and energy as well as for fire safety. In some regions, hurricane preparation offers another possibility. A schedule of maintenance tasks and a shopping list of tools and supplies provide home owners with practical help and reasons for referring to the home owner guide.

Illustrations

You can include digital photos to add interest and to further clarify details. Tasteful clip art offers another method to create visual interest. Some guides incorporate icons such as a light bulb to designate a good idea or short cut and a caution sign to call attention to a safety warning.

Positive Wording

Eliminate unnecessary negativity with judicious editing. For instance the repeated warning "[Builder] will repair <type of cosmetic damage> only if it is noted during the orientation" can be reworded as follows: "We will confirm the good condition of <component> during the orientation. [Builder] will address any items noted. Following that inspection, fixing cosmetic damage will be your responsibility."

Conversational Tone

Compose this material as if you are talking directly to your home owner. Saying "It is recommended. . . ." distances the company from its home buyers. _Who_ recommends? Simply say "We recommend. . . ." Not only is this approach friendlier, the home buyer knows who you are anyway.

What's Important?

Everything in the home owner guide—at one point or another—is important. Saying about a particular point that "It is important to remember. . . ." or worse, "It is very important to remember. . . ." suggests that other points are unimportant. Get straight to the point: "Remember to remove hoses from outside faucets before the first freeze to prevent burst pipes."

Emergency Tips

Builders often present tips for emergencies (for instance, lack of heat, hot water, or electricity; a major plumbing leak) in one section. But few home owners will have emergencies with all of these items at once. These pointers will prove easier for home owners and others to find if they are listed under the component title (see Figure 3.4). If finding information in your home owner guide is more difficult than calling the warranty office, home owners are more likely to call the warranty office. One of the normal maintenance points could be as useful in an emergency as the troubleshooting tips. Separation of the information increases the possibility of the home owner missing valuable instructions, suggestions, and ideas.

Appropriate Disclaimers

Builders, quite fairly, exclude some items from warranty coverage because they have little or no control over the contributing factors. These items include alarm systems and smoke detectors serving their intended purposes, possible visibility of paint touch ups, and dye-lot variations in replacement of tile or carpet. Express these disclaimers clearly as in Figures 3.6 and 3.7.

FIGURE
3.6 Home Owner Guide Entry: Alarm System

Homeowner Use and Maintenance Guidelines

If your home selections included wiring for an alarm system to be activated later, you will arrange for the final connection and activation after you move in. The alarm company will demonstrate the system, instruct you in its use, and provide identification codes for your family. We recommend that you test the system each month.

[Builder] Limited Warranty Guidelines

[Builder] will correct wiring that does not perform as intended for the alarm system. **[Builder] makes no representation that the alarm system will provide the protection for which it is installed or intended.**

FIGURE
3.7 Home Owner Guide Entry: Warranty on Drywall

[Builder] Limited Warranty Guidelines

During the orientation, we confirm that wall surfaces are in acceptable condition.

Lighting Conditions

[Builder] does not repair drywall flaws that are only visible under particular lighting conditions.

One-Time Repairs

One time during the materials and workmanship warranty, [Builder] will repair drywall cracks and nail pops and touch up the repaired area using the same paint color that was on the wall when the home was delivered. Because of the effects of time on paint and wallpaper, as well as possible dye-lot variations, touch-ups are unlikely to match the surrounding area. Repainting the entire wall or the entire room to correct this situation is your choice and responsibility. You are also responsible for custom paint colors or wallpaper that has been applied subsequent to closing.

Related Warranty Repairs

If your home needs a drywall repair because of poor workmanship (such as blisters in tape) or other warranty-based repair (such as a plumbing leak), [Builder] completes the repair by touching up the repaired area with the same paint that was on the wall when the home was delivered. If more than one-third of the wall is involved, we will repaint the wall corner to corner. You are responsible for custom paint colors or wallpaper that has been applied subsequent to closing. Because of the effects of time on paint and wallpaper, as well as possible dye-lot variations, touch-up will vary from the surrounding area.

Gaining Home Buyer Acceptance of Guidelines

To decrease conflicts and increase customer satisfaction (translation, repeat, and referral sales), builders need to work hard helping customers understand the quality they are buying and how the warranty backs up that quality. Home owner maintenance responsibilities are equally important, and the builder needs to incorporate them into buyer education.

Builders sometimes forfeit reputations (and sales) when they assume that home buyers truly see the company's quality in model homes and that they read the home owner guide cover to cover. Handing the home buyers the guide and saying, "This is important, please read it," fails to accomplish the goal. However, team work manages the task effectively.

While repetition is the key, it can become annoying unless you use a variety of approaches. The examples that follow offer illustrations of how all frontline personnel can play a part. With a bit of effort from each person who works with home buyers, your company can improve customer satisfaction.

Start at the Beginning

Include a brief and accurate overview of limited warranty coverage and the structure of warranty service as part of the sales presentation. Encourage this conversation with a flyer or poster similar to that shown in Figure 3.8.

Perfection eludes all builders in all price points on some details; the natural properties of materials in a new home are not negatives, they are simply reality. Calling attention to examples such as variations in wood grain, tile colors, and so on sends a message that the home buyer can trust the company. Today's consumer appreciates this level of candor. However, reading and hearing about such differences simply does not have the same impact as seeing them does.

Before completing the purchase agreement, seasoned sales professionals who are committed to earning referrals walk the model home, inside and out—including the garage and basement—with buyers. They take time to encourage home buyers to look at and focus on the quality of concrete flatwork (if a model driveway has a $\frac{1}{8}$-inch crack, point it out), exterior trim (typically rougher than interior trim), carpet seams (which go unnoticed because furnishings distract the eye), interior trim details (notice some nail heads are visible when you look closely), and so on. Thus, the salesperson forestalls later disappointments and makes referrals more likely. Just as valuable, customers conclude they are dealing with forthright, honest people.

Salespeople can contribute still further to buyer expectations about warranty and maintenance when they review (with the home buyer) the bulleted summaries at the beginning of each section of the home owner guide. Figure 3.9 offers an example of a potential bulleted summary of a "Caring for Your Home" section.

Continued Emphasis on Product Realities

Selection coordinators also need to call attention to product realities (cherry cabinets darken over time) and maintenance responsibilities listed in the home owner guide (marble flooring needs regular care to look good), especially

A tall salesman in a plaid shirt, who was sporting a handlebar mustache, approached a custom home buyer while she was reviewing double ovens at an appliance showroom. Based on his appearance, she concluded that he was a classic, fast-talking salesman. She responded to his, "How can I help you today?" with the equally classic customer attitude, "I'm just looking."

He persisted, "What appliance are you interested in?" Reluctantly she responded, "Double oven." Then he won her over with this commitment: "If you'll give me five or six minutes, I'll show you double ovens at three different price points and explain the advantages and disadvantages of each."

The customer was so impressed with the man's forthright approach that she selected the most expensive double oven in the store and has returned repeatedly to buy other appliances. In talking with her builder, she said, "It was so refreshing to find someone who tells you the good and the bad. You know you're hearing the truth when the all the realities are mentioned."

FIGURE
3.8   Flyer to Generate Interest in Warranty at the Sales Office

Please . . .
Ask us about our
Limited Warranty!

Why Do We Want to Tell You about Our Limited Warranty?

In the midst of decisions about floor plans and carpet colors, cabinet and tile selections, light fixture and appliance choices, who wants to think about home maintenance and warranties? When the next task is to set appointments with the design center for selections and the superintendent for your preconstruction meeting, who wants to think about repair appointments?

We do—because we

RESPECT your right to make an informed purchase. We volunteer information even when you are too excited to ask for it.

WANT you to be satisfied long after you move into your new home. We help you understand the separation between warranty and maintenance: our job and your job.

BELIEVE that an informed buyer enjoys the building process and the new home more. We work hard to anticipate items that might surprise you and make a sincere effort to talk about them sooner rather than later.

HOPE that nothing will ever go wrong in your new home and work hard toward that end. Still, we are realistic enough to know that something might need repair. We have a system to respond when you need us, a system that works much better if you are informed about how to use it to obtain the warranty service you deserve.

SO, we invite you to ask us about our Limited Warranty.

For complete details, review a copy of the limited warranty and the "Caring for Your Home" section of your [Builder] Home Owner Guide.

FIGURE

3.9   Home Owner Guide Entry: Bullets Summarizing Caring for Your Home

- Homeowner Use and Maintenance Guidelines—introduction to the maintenance information in this manual
- [Builder] Limited Warranty Guideline—introduction to the criteria [Builder] uses to screen warranty items
- Warranty Reporting Procedures—standard, emergency, miscellaneous, and appliance warranty procedures
- Warranty Item Processing Procedures—a simple description of a complex process
- Help Us to Serve You—what you need to know so we can provide effective warranty service
- Warranty Service Summary—a one-page guide to the contact person in various service situations
- Fire Prevention—reminders to prevent fire in your home
- Extended Absences—tips for preparing for absences and reminders for the day you leave
- Energy and Water Conservation—suggestions for consuming energy and water wisely
- Appliance Service—a worksheet for recording serial and model numbers along with manufacturer service phone numbers
- Home Care Supplies—create a shopping list of tools and supplies you need to care for your home.
- Maintenance Schedule—make notes about routine maintenance tasks and plan your schedule
- Air-Conditioning Through Wood Trim—an alphabetical list of the items in your home, including maintenance hints, warranty criteria, and troubleshooting tips
- Forms—for your convenience when reporting warranty items and giving us feedback about this manual

when a home buyer is choosing an unfamiliar finish material. Use a positive approach similar to the following example:

> We want you to make informed decisions. If you have not had white marble entry tile before, it's a good idea to review the information in our guide. This entry outlines the maintenance this product will need as well as our warranty commitment to you on the product.

Familiarity with products and colors shown in models can be another valuable tool. "You can see the variation in how different species of wood take stain in our Wellington model. Notice the stair rail compared to the hardwood flooring in the entry."

During Construction

Construction personnel play a part in this home buyer education effort as well. For instance, they can point out some items at the preconstruction meeting and the frame stage tour.

Preconstruction Meeting. The agenda for this meeting should include comments from the construction superintendent on your company's quality inspection system. Emphasize that regular inspections the company performs reduce warranty items later and benefit both the home buyers and the company. Let home buyers know that the person leading the frame stage tour will point out some of the high-quality building techniques used inside the walls of their new homes.

Frame Stage Tour. When a company representative takes a home buyer on a tour of the customer's home at frame stage, he or she can call attention to details that will be out of sight when the home is complete. The conversation should include warranty and maintenance information to show the division of responsibilities. If your framers plane and shim studs before the drywall is hung see that this fact is on the agenda with the explanation that this procedure helps to ensure plumb walls. Talking about drywall serves as an obvious lead into a conversation about nail pops and drywall separations. Your company rep can briefly review the warranty service on these items. (Typically they are fixed one time, and the company usually recommends that the repairs take place at the end of the warranty period.) The rep can also mention here the wisdom of waiting to apply custom paint colors or wallpaper until after this service has occurred.

Likewise, by calling attention to places where plumbers protect pipes from nails with metal plates, the company rep can easily raise the subject of interior moisture levels and the home owner's responsibility to use the home's ventilation systems and conduct routine inspections for signs of water intrusion.

The rep describes the insulation that will soon be installed and cautions the home buyer not to disturb attic insulation, which would reduce its effectiveness. And again the rep connects something the company does with the home buyer's maintenance responsibilities after move in. The message is that the builder does some things, and then the home owner takes over. To reinforce the information at a couple of points in such dialogs, the rep opens the home owner guide and shows the home buyer where to find a discussion of these details in writing.

At Delivery

You can review many maintenance and warranty details with the home buyer at orientation. Keep the tone positive, begin such discussions with what the company fixes, then describe the home owner responsibilities. For example, while demonstrating window operation, the builder's representative might explain the following information about internal moisture:

> If you notice condensation between the two panes of glass, contact our warranty office. We provide warranty coverage if a window lock fails or if the windows are difficult to open and close. If you notice condensation on the inside of the window, it comes from moisture in the home generated by cooking, showers, laundry, and so on. Because those activities are outside of our control, our warranty excludes that condition. Our home owner guide provides some suggestions on using your home's ventilation systems effectively to prevent condensation. You'll find them here on page <page number>. (Show the home buyer the entry in the guide.)

You can approach nearly every component in a home with this type of description. Some items take a couple of minutes, others just a few seconds. Certainly, you would include warranty reporting procedures on the agenda for a home tour. Turn to the guide and call attention to an explanation of warranty decisions similar to the one in Figure 3.10. Discuss with warranty personnel which components need special emphasis. When you open the home owner guide, go to these topics and show the home buyers where to find further details.

Home Buyer Pre-Move-In Seminar

Chapter 1 includes a discussion of home buyer seminars that mid- to large-volume builders can use as models to develop a pre-move-in seminar for home buyers. The scope of material covered in such a program might include the following items:

* Describe your steps to prepare the home for delivery.
* Point out details about the orientation. (Ask participants to turn to the related pages in their home owner guides.)
* Review the buyers' tasks in preparing for closing (again, listed in the guide).

FIGURE

3.10 Home Owner Guide Entry: Warranty Coverage Decisions

When we receive a warranty service request, we may contact you for an inspection appointment. Warranty inspection appointments are available Monday through Friday, 7 a.m. to 4 p.m. We inspect the items listed in your written request to confirm warranty coverage and determine appropriate action. Generally, reported items fall into one of three categories:

* trade contractor item
* in-house item
* home maintenance item

If we require a trade contractor or an in-house employee to perform repairs, we issue a warranty work order describing the situation to be addressed. If the item is a home maintenance matter, we will review the maintenance steps with you and offer whatever informational assistance we can. Occasionally the inspection step is unnecessary. In that case, we issue the needed work orders and notify you that we have done so.

Our criteria for qualifying warranty repairs are based on typical industry practices in our region and meet or exceed those practices. Please note that we reserve the right, at our discretion, to exceed these guidelines if common sense or individual circumstances make that appropriate, without being obligated to exceed all guidelines to a similar degree or for other home owners whose circumstances are different.

- Demonstrate how easily home buyers can find answers to routine questions or guidance in an emergency by turning to the guide's alphabetically arranged information on components in the guide.
- Explain warranty reporting procedures for both emergency and nonemergency warranty items.

After-Move-In Warranty Meetings

Some builders believe that home buyers have too much on their minds to remember all the detailed information typically covered in a pre-closing orientation. These companies have created two separate meetings: a shortened pre-closing orientation followed a few weeks later by an after-move-in warranty and maintenance meeting. The abbreviated orientations cover critical items such as emergency shutoffs and operation of the heating, ventilating, and air-conditioning (HVAC) system. The builder's rep and the home buyers also confirm the acceptable condition of all cosmetic surfaces. Then, within two to four weeks of closing, a warranty rep meets with the home owners now living in their new home.

Sitting at the kitchen table or in the family room, they review maintenance and warranty information and regularly refer to the home owner guide during the discussion. The rep tours the home with the buyers, demonstrates the details of the operating systems, and discusses how to care for surfaces. Often by this time, the home owner has noted a few items so the warranty rep instructs the home owner in compiling a warranty list for repair orders. You can easily blend this approach with the builder-initiated warranty visit concept by setting up this appointment at the end of the orientation.

When the home owner keeps this appointment and the rep conducts an effective meeting, this approach works well. However, concerns to watch for include home owners who cancel the appointment and never participate in this review. Another is the possibility that a home owner will experience difficulty with one of the home's features or damage something out of ignorance prior to this follow-up meeting. All procedures come with pluses and minuses; this one is no exception.

Online Warranty Request

When home owners use online systems to report warranty items, builders can take advantage of yet another opportunity to point out warranty guidelines. To establish a credible system, you have to ensure that the criteria and the repair described critically match those the home owner received on paper. Figure 3.11 provides an example of this approach.

Warranty Visits

When warranty reps inspect homes they should carry the home owner guide with them. The conversation can be as simple as "This is a maintenance item. Page <page number> lists the steps for taking care of <item>. I'll be glad to go

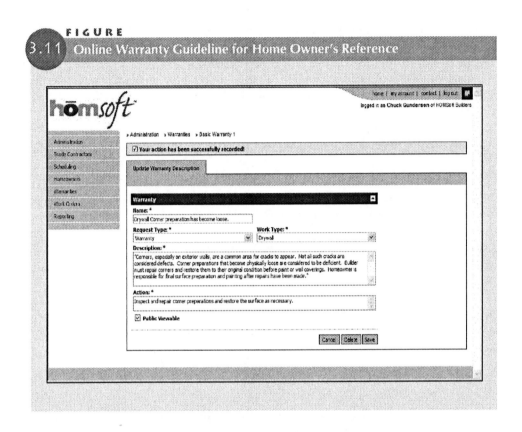

FIGURE 3.11 Online Warranty Guideline for Home Owner's Reference

through the steps with you if that would be useful." Note that when the home owner responds with "Yes, I read that, but let me tell you what led up to this problem. . . ." the warranty rep faces a circumstantial judgment situation and needs to listen carefully to the customer's point of view. An exception may be appropriate. (See Chapter 5 for more on such judgment calls.)

Confirming Letters

When you need to enforce procedural details, outline a complex repair, or deny a home owner's request for a repair, confirm that conversation with a letter that refers to the related page or pages of your home owner guide. For the home owner's convenience, you can include a copy of the pages you mention.

Interpreting and Applying the Guidelines

With a product so personal and so significant financially, buyers are certain to have strong opinions. A home owner's view of what a builder should correct or how to perform a particular repair sometimes differs from the builder's warranty commitment. This situation seldom creates a problem when the builder wants to do more than the home owner requested, but it can become a serious problem when the builder intends to do less. The company greatly reduces this inevitable friction when its employees adhere to the following practices:

- Align expectations as described in Chapter 1.
- Practice quality management aggressively during construction.
- Deliver homes that are complete and clean.
- Plan and carry out proactive warranty contact with home owners after move-in.
- Respond promptly to random warranty requests between standard checkpoints.

Industry Standards

Warranty personnel may defend a condition by saying, "That meets industry standards." The home owner's response is likely to be fueled by anger. The home owner did not buy industry standards. The home owner purchased the builder's marketing promises: craftsmanship, attention to detail, sophisticated styling—in other words, the advertising claims about the company's quality. And none of them say, "We build to industry standards."

For you to use this phrase after the home buyer has paid for the home can feel to the customer like a betrayal. Warranty personnel will experience somewhat better responses from home owners if they use the builder's name: "That meets [Builder] standards; this item appears this way in the models."

Disagreements

In spite of these worthy efforts you should expect occasional disagreements. One type involves a guideline you provided from the beginning. Sometimes when a customer *reads* that "Separations of $^3/_{16}$ inch or greater in concrete flatwork" will be caulked, that measurement seems acceptable. When the same customer *sees* a separation of $^3/_{16}$ inch in her driveway, it is much less acceptable.

Another type of disagreement involves a condition for which you did not provide a written guideline—perhaps because none exists. Such items are often subjective (carpet seams, paint coverage, or wood grain variations). While the first type of disagreement is somewhat easier to resolve because you documented the company's warranty commitment as part of the contract with the home buyers, neither type is desirable.

To avoid as many disagreements as possible, take note of conditions that cause difficulty repeatedly. Review the "teachable moment" suggestions in Chapter 1 to identify an effective method of educating home buyers about standards for the item. If this study leads to a written document designed to highlight information, ensure that the information is consistent with the home owner guide. Reproduce the entry exactly as it appears in your guide and add a statement referencing the page number (Figure 3.12).

Training Activities

The critical warranty service training should be ongoing and involve all frontline staff. Integrating the home owner guide into conversations with home buyers needs to become a natural part of everyone's workday. Role plays should involve

FIGURE

3.12   Example of Highlighting: Wood Flooring

To ensure that you make informed decisions, we ask that you take note of these details that appear on page <page number> of your [Builder] Home Owner Guide.

Hardwood Floors

The warmth and ambiance of wood floors add something special to any decor. However, wood floors are not impervious to normal wear, and we want you to know what to expect before making this choice for your home. Please review the following information and keep in mind that preventive maintenance is the primary goal in caring for hardwood floors.

Cleaning. Sweep on a daily basis or as needed. Never wet mop a hardwood floor. Excessive water causes wood to expand, possibly damaging the floor. When polyurethane finishes become soiled, damp-mop with a mixture of one cup vinegar to one gallon of warm water. When damp-mopping, remove all excess water from the mop. Check with the hardwood floor company for cleaning recommendations if your floor has a water-based finish.

Dimples. Heavy furniture or dropping heavy or sharp objects can result in dimples.

Filmy Appearance. A white, filmy appearance results from moisture, often from wet shoes or boots.

Furniture Legs. Install floor protectors on furniture used on hardwood floors to allow chairs to move easily over the floor with less scuffing. Clean the protectors regularly to remove grit that may accumulate.

Humidity. Wood floors respond noticeably to changes in humidity in your home, especially during winter months. A humidifier helps, but does not completely eliminate this reaction.

Mats and Area Rugs. Use protective mats at the exterior doors to help prevent sand and grit from getting on the floor. These substances are wood flooring's worst enemy. However, be aware that rubber backing on area rugs or mats can cause yellowing and warping of the floor surface.

Recoat. If your floors have a polyurethane finish, in six months to one year you may want to have an extra coat of polyurethane applied. You should have a qualified trade contractor apply this extra coat. The exact timing will depend on your particular lifestyle. If another finish was used, please refer to the manufacturer's recommendations.

Separation. Expect some shrinkage around heat vents or any heat-producing appliances.

Shoes. Keep high heels in good repair. Heels that have lost their protective cap (thus exposing the fastening nails) will exert thousands of pounds of pressure per square inch on the floor. They will mark your wood floors.

Spills. Clean up food spills immediately with a dry cloth. Use a vinegar and warm water solution for tough spills.

Splinters. While floors are new, small splinters of wood can appear.

Traffic Paths. A dulling of the finish is likely to occur in heavy traffic areas.

Warping. Warping will occur if the floor becomes wet repeatedly or is thoroughly soaked even one time. Slight warping in the area of heat vents or heat-producing appliances is also typical.

Wax. Waxing and the use of wood-cleaning soaps and similar products are neither necessary nor recommended. Once you wax or clean a polyurethane finish floor with such a product, recoating is difficult because the new finish will not bond to the wax.

I/we have read the information regarding care and performance of wood floors. We accept the responsibilities that come with the beauty of wood floors.

_____   _____   _____   _____
Purchaser   Date   Purchaser   Date

sales, selections, construction, and warranty personnel. For each group, focus on the typical conversations each staff member in that group might have with home buyers or home owners. Warranty personnel can provide suggestions as to which procedures, components, and conditions your company needs to emphasize.

Review warranty and maintenance information in detail, one or two components at a time. Frontline personnel can present or defend a guideline more effectively if they understand the reasons for it. Listen to concerns, questions, and suggestions for improvements where needed and keep in mind that standards and guidelines evolve over time.

Working with the warranty staff, review decision-making guidelines and responses to predictable situations such as those listed in Chapter 5. Emphasize how and when exceptions to normal guidelines might be appropriate and when to check with management before exceeding normal boundaries. Explore case-study situations that provide your staff with judgment-call practice. Frontline warranty professionals must know the written guidelines as a starting point only. They also must know how to handle unusual situations. They must take circumstances into account and apply common sense to every situation. Judgment skills, as discussed in Chapter 5, "Decisions," help. Ultimately, builders need to analyze and resolve each situation case by case to arrive at fair decisions.

Warranty Service Structure

W arranty activities involve hundreds of details from major to minor, from practical to bureaucratic, from objective to emotional. Because homes and the people who live in them are richly varied, the issues that arise include unending twists and surprises. Add trade contractors and weather, and you have a real puzzle.

Trying to manage this dynamic mass without a plan is certain to lead to difficulties. To gain control, begin by deciding on a structure for your warranty service. This structure provides a framework for receiving and processing warranty items. Core elements include the traditional four steps of consumer complaint handling.

Input. You learn of the item(s) in need of attention (inspection and repair or review of the home owner's maintenance responsibilities). Various methods of managing this step are detailed in this chapter.

Analysis. Discussed in Chapter 5 analysis involves deciding whether an item should be fixed and, if so, who should do the fixing.

Response. Notwithstanding the occasional compromise, the choices usually come down to providing the requested service or denying it. Examples of both are illustrated in Chapters 6 and 9.

Follow-Through. The last step ensures that work you ordered is completed appropriately and that service denials are documented. Chapters 8, 10, and 11 address several aspects of follow-though.

In addition, plan transitional communication and services for home owners as they enter the warranty period and again as they exit from warranty upon the expiration date. You will also need policies and procedures for emergencies and other situations that are discussed later in this chapter. (Still more specialized procedures are addressed in Chapter 7.)

Transition into Warranty

By the time the closing date arrives, most home buyers have established relationships with a salesperson, mortgage loan officer, design consultant, and construction superintendent. Although various hand-offs have occurred along the way,

---

Take steps to establish trust and confidence.

the customers remained somewhat in contact with the previous individuals at the same time a new professional joined the team.

Now they experience yet another hand-off, this time to the warranty department. This hand-off is different from the earlier ones in several ways. Culminating a decision made weeks or months earlier, the buyers are about to pay for the home. Turning back will no longer be an option. The leverage of "I'm not closing unless. . . ." will be gone. This finality can make some home buyers nervous.

Additionally, most builders prefer that after closing, contact with personnel who handled the early phases of the new home experience stop or at least diminish dramatically. (This tradition is changing—see Chapter 13.) Buyers find themselves getting to know and trust a new personality from warranty at the same time that established relationships come to what can feel like an abrupt end. Add the work of moving, and you can easily see why home buyers can be a bit volatile. A smooth transition to warranty essentially maintains your customers' comfort level and satisfaction.

Hand-Off Communication

Of all the hand-offs in the new home experience, this hand-off to the warranty staff is one of the most significant. Ensure that background information from sales, selections, and construction is shared with the warranty staff. If the warranty rep attends some of the community team meetings (Chapter 1), much of this transfer of information already will have occurred.

Early Introduction

One of the easiest methods of building rapport between the home buyers and their warranty reps is creating an opportunity for them to have contact prior to the closing. Make this your first transition service objective. Many methods work, and you may use several of the following methods to build scheduling flexibility into the process:

- The warranty rep can stop by for an introduction during the frame stage tour.
- The agenda for a preclosing home buyer seminar can include a "Meet Your Warranty Team" segment.
- The warranty rep might make a phone call to the home buyers a few weeks prior to closing "just to introduce myself."
- At the least, the salesperson could give the buyers the warranty rep's business card along with some information that establishes the warranty rep's credentials.
- A more significant opportunity occurs if the warranty rep conducts the orientation. The superintendent, with or without the salesperson present, might begin the orientation and then turn the home and the home buyers over to the warranty rep. Some builders have the superintendent sign his or her portion of the orientation agenda and hand the file over to the warranty rep to symbolize the hand-off in a tangible way.

* If the superintendent conducts the orientation, the warranty rep can stop in toward the end, meet the home buyers and provide an overview of warranty procedures.

Service Continues

The second transition service objective is to ease the customer off of the excitement and intensity of the building process. Home owners often complain that once their builder gets paid, attention ceases. Survey statistics typically show a drop in customer satisfaction of 8 to 11% between closing and the end of the first year. Researchers attribute this decline largely to the buyers feeling abandoned. You can prevent much of this loss in satisfaction (see more discussion on this vital topic in Chapter 13). Your transition procedures demonstrate that customer service and appreciation continue after you've collected the final payment. As an added benefit of transition efforts, any developing issues are discovered quickly so you can work to resolve them.

Plan a minimum of three contacts during the first month after move-in, including contact with the warranty staff as well as with other departments. The result should be overlapping attention from several individuals. Suggested procedures follow, and you will undoubtedly think of many others:

* Regardless of who conducts the orientation, consider setting up the first warranty visit appointment at the end of the orientation. A detailed discussion of builder-initiated warranty visits appears later in this chapter.
* Send a welcome-to-warranty letter the week after closing. Figure 4.1 offers an example that confirms the builder-initiated warranty visit.
* The superintendent might meet with the home owner a few days after move-in to confirm that orientation items are complete, check on satisfaction, and answer questions.
* The warranty administrator can call the home owners shortly after move in. "I'm just checking to see how your move is going and whether you have any questions."
* The warranty rep can visit the home owner two to four weeks after move in and review utility shutoff locations, warranty coverages, key maintenance points, and reporting procedures (see "After-Move-in Warranty Meetings" in Chapter 3 for more information on this idea).
* The salesperson might visit the home owners within the first two weeks of their move-in with a "Welcome to the Community" gift. This visit provides the company with another opportunity to check on satisfaction, answer questions, and ask for referrals.
* The selections consultant may be involved in transition services. Seeing the buyers' choices completely installed can be useful to the design consultant, and simultaneously, home owners get transition attention. A gift-basket approach that includes coupons and samples of suitable home care products adds extra value.

FIGURE
4.1   Welcome to Warranty Letter (traditional approach)

Dear <Home Owner>:

On behalf of [Builder] I'd like to welcome you to <community> once again. We all hope that your move went well and that you are enjoying your new home.

While we believe that we delivered an excellent home to you, we recognize some items in the home may require follow-up work. Our limited warranty spells out the services we provide in this regard.

If you notice any nonemergency warranty items that need attention, please note them on a warranty service request form. You will find these in your [Builder] Home Owner Guide at the back of Section 8, "Caring for Your Home." You are welcome to mail or fax the form to the warranty office. If you prefer, visit our Web site: <Web site address> and click on "Warranty Request" to send in your request.

Emergencies are rare, but if one occurs please call our warranty office during normal business hours, Monday through Friday, 8:00 a.m. to 5:00 p.m. In the event of an after-hours emergency, refer to the Emergency Phone List sticker inside your kitchen cabinet. Please remember to call our office on the next business day so that we can document the item for your file and follow up with you.

You can find complete details about our warranty procedures and guidelines in your [builder] Home Owner Guide, Section 8, "Caring for Your Home."

Please feel free to call me if you have any questions. I look forward to working with you in the coming months.

Sincerely yours,

Warranty Manager
[Builder]

● Host a "Meet Your Neighbor" reception. Depending on volume, your company might sponsor this event monthly or quarterly, and depending on the number of potential attendees, you might invite home owners to bring a friend. Builders often find viable prospects this way.

   Note that you don't need to use all of these ideas. Select those likely to appeal to your home buyers. Use a variety of methods and personnel. Balance your transition activities to achieve your objectives without becoming too intrusive and annoying busy families.

Input Options

A successful transition brings the home owner smoothly into warranty. Next, you must prepare for the routine processing of warranty items, beginning with input. Heated debates occur among builders about the timing and frequency of reports of warranty items. Choices to consider include the following questions you need to answer:

* Should you encourage home owners to report items directly to the trades?
* Should you use random reporting or establish checkpoints for consolidated lists?
* If you create routine checkpoints to receive consolidated lists, how many should you plan and when should they occur?
* Is a midyear checkpoint beneficial?
* Should you remind the home owner of the approaching expiration of the warranty?
* With routine checkpoints, who should initiate the communication—you or the home owner?
* How will your company and your trades handle emergencies, especially after hours and on weekends?

And while your service can be world class, you should be prepared for complaints and questions.

Once you have made choices about the basic system, you are likely to need policies for responding to warranty requests that home owners submit outside of the routine methods. For example, consider the three typical issues that follow:

* How will you respond to home owners who report items between the routine checkpoints?
* And what will you tell home owners who submit overlapping lists?
* How should personnel outside the warranty department respond when home owners want to report warranty items, for instance to their salesperson or superintendent?

Home Owner Contact with Trade Contractors

Resist the temptation (which may be strong for small-volume builders) to give home owners lists of phone numbers and instructions to call trade contractors directly and involve the builder only if a disagreement develops. To maintain the desired levels of quality, you must be aware of the number and nature of warranty items addressed. Furthermore, hearing from a home owner only after a problem with a trade contractor has occurred can cause damage to both relationships. Your home owners' direct contact with trades is acceptable for after-hours emergencies as described later in this chapter or minor contacts regarding appointments resulting work you have ordered.

Sometimes, what appears on the surface to be better service turns out to be just the opposite.

Random Reporting

Some builders tell home owners to submit items one or more at a time whenever they notice them. Proponents of this system believe that home owners report fewer warranty items this way than when the home owner is asked to make a consolidated list at a set time. Other companies use random reporting because they believe it is more service oriented. They ask, "Why should a home owner have to wait for service if something is not right?" The long-term impact of the random approach can be inconvenient for home owners who repeatedly find themselves taking time for warranty appointments.

This method is also highly inefficient for the trades and the warranty personnel. Repeated trips to homes for inspections and repair appointments can result in slower service for everyone because so much time is invested in going to and from the same homes for one or two items at a time. This process is comparable to making a separate trip to the grocery store for each item on your list rather than buying everything at once.

Routine checkpoints (which are optional for the home owner) along with an hospitable response to miscellaneous reports between those checkpoints combines the best of both systems and benefits for the home owners, your trades, and your warranty personnel.

Routine Checkpoints

Once you commit to offering routine checkpoints, you face decisions about how many and when they should occur.

First Checkpoint—30, 60, or 90 Days?

Experience shows that 90 days is too long; home owners on that schedule frequently contact their builder with items between 45 and 60 days because they are anxious for corrections. Thirty days may be too soon if most households in your communities have both adults working—they are barely settled within that time. Offering the customer a choice between 30 and 60 days seems to work well, and it still retains the efficiencies that consolidated requests offer to all parties.

Provide home owners with several methods to document their warranty requests. If possible, include an online method of reporting items similar to the one in Figure 4.2 and traditional warranty service request forms such as the one in Figure 4.3. Include at least two of these in the home owner guide. This form gives them a convenient place to note items or questions and helps standardize the information received in the warranty office and thus makes processing easier and faster.

Midyear Checkpoint

A few builders schedule a five- or six-month warranty visit as one of their planned contacts. To some home owners this plan suggests that the builder expects a high

FIGURE
4.2 On-Line Warranty Request

a. Opening Screen

The home owner clicks on the word *contact* in upper right, the screen in Figure 4.2b appears, and the home owner enters warrant request items.

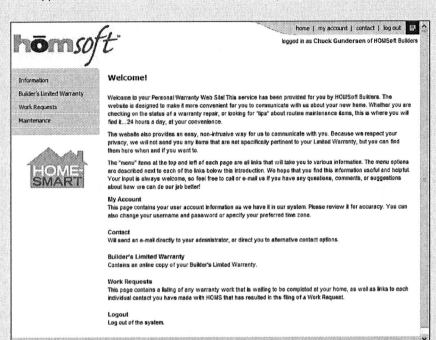

b. Second Screen

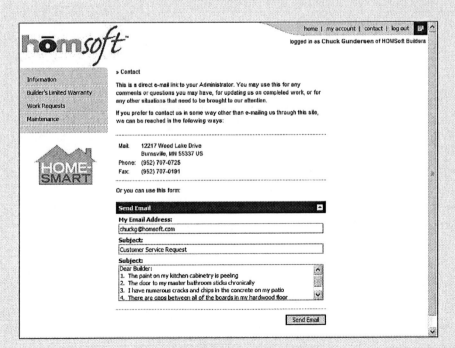

FIGURE

4.3 Warranty Service Request Form

[Logo]   # Warranty Service Request

For your protection and to allow efficient operations, our warranty service system is based on your written report of nonemergency items. Please use this form to notify us of warranty items by mailing this request to the address listed above or fax it to 555.555.5555. You can also send in a warranty request by clicking on the *Warranty Request* button on our Web site: www.builderwebsite.com. We will contact you to set up an inspection appointment. Service appointments are available from 7:00 a.m. to 4:00 p.m., Monday through Friday. *Thank you for your cooperation.*

Home Owner ___*Cindy Buyer*___   Date ___*7/14/20–*___

Address ___*3399 West Port Ave*___   Community ___*Bailey's Cove*___

Home phone ___*555-555-1234*___   Home site no. ___*3-27*___

Work phone _____   Plan ___*Trinity Bay*___

Cell phone ___*555-702-0050*___   Closing date ___*4/3/20–*___

	Service Action		
Service Requested	Warranty*	Courtesy*	Maintenance†
We are getting water in the northwest corner of our			
basement. Please help!			

Comments

I'm home most mornings but will arrange my schedule
to meet with someone any time. We must get this
resolved so we can finish the basement!!!

---

* The term *Warranty* or *Courtesy* indicates [Builder] will address.
† *Maintenance* indicates a home owner responsibility.

Home Owner ___*Cindy Buyer*___

number of warranty items to arise. Busy home owners may be annoyed by what can feel like a constant need to take time off of work, first to meet the warranty rep and second to provide access for repairs to occur.

Plan a level of attention equal to your service capacity so that you can executive your service commitments consistently and effectively.

Because of the extra appointments, warranty staff and trades' personnel are operating at lower efficiency. Planning a third contact adds significantly to warranty staff workload, and often results in these promised appointments occurring late or not at all. Certainly you can add more personnel to avoid getting behind, but this response can increase home prices, so proceed with caution. Customer satisfaction is greater when builders promise a bit less and do it well than when they promise a lot, and fail to deliver.

Builders often feel pressured to include a midyear warranty visit because their third-party survey company conducts a midyear survey. Especially considering that research shows customers are oversurveyed, ask that this midyear survey be eliminated. Spend what you save serving customers instead of annoying them with yet another survey. Frederick C. Reichheld covers this issue in his book, *The Ultimate Question*, published by the Harvard Business School Press.

Warranty Expiration: Year-End Checkpoint

More debate occurs when the topic of warranty expiration comes up. Again, some builders believe that, if they let the expiration of the warranty pass without comment, they will have less warranty work to do. However, the experience of builders who contact home owners before their warranties expire actually shows that the overall volume of work per home remains the same. A routine year-end contact generates all the same benefits that were listed for a 30- or 60-day visit. Additionally, by controlling the timing of year-end warranty visits, you can eliminate most disagreements with trades about their warranty liability. Finally, your name is on this home; you want it to impress visitors (some of whom may be in the market for a new home). Depending on how proactive you choose to be, the end of warranty can be handled in one of several ways as follows:

- Title one of the warranty service request forms you include in the home owner guide, "Year-End Warranty Request."
- Send a letter, postcard (with first-class stamp), or e-mail reminding the home owner of the approaching expiration and asking the home owner to submit warranty requests prior to the anniversary date of their closing. A sample end of warranty letter appears in Figure 4.4.
- Call to remind the home owner and set up an inspection appointment during the same conversation, if appropriate. If you are unable to reach the home owner or fail to get a return call within two business days, follow up in writing.

Once you have your warranty service structure in mind, consider implementing what many see as the current best practice for new home warranty:

FIGURE

4.4   Year-End Warranty Letter

Dear <Home Owner>:

Over ten months have gone by since you closed on your new [Builder] home. We hope you have found your home and the surrounding community to be a pleasant and comfortable place to live.

As you are aware, the Materials and Workmanship portion of your [Builder] Limited Warranty will expire on

_____<date>_____

If you have noticed any items that require warranty attention, please fill out a Warranty Service Request and mail or fax it to our office by <date>. If you prefer, visit our Web site: <Web site address> and click on "Warranty Request" to send in your request.

Upon receipt of your request, we will contact you to set an appointment to view items you have listed and answer questions you may have. If you have no items, but would like us to check the home, please call our office to arrange an appointment.

For additional information about warranty coverage please refer to your [Builder] Home Owner Guide, Section 8, "Caring for Your Home."

After the expiration of your Materials and Workmanship portion of your [Builder] Limited Warranty, the Structural coverage remains in effect as we remain available to answer questions you may have about maintenance of your home.

Sincerely yours,

Warranty Manager,
[Builder]

builder-initiated warranty visits. With this procedure, you take a more proactive approach rather than putting warranty reporting responsibility on the home owners alone.

Builder-Initiated Warranty Visits

Asking home owners to submit a consolidated list (for instance, 30 or 60 days after move in) seems to be logical and efficient. It is neither. Savvy builders ask to visit the home to confirm that all components are functioning as the builder intended. Inspecting any items the home owner has noticed is a logical part of these appointments as follows:

* As one of a series of customer meetings (see Chapter 1) builder-initiated warranty visits continue familiar procedures.
* The message that builder-initiated warranty visits send to home owners is loud and clear: "We stand behind our homes."
* Knowing that they have planned meetings coming (at 30 or 60 days and again at year end), fewer home owners contact the warranty office during their first weeks in the home and between routine visits.
* These visits create an opportunity to confirm that the home owner is maintaining drainage correctly. If the warranty rep notes yard additions (such as fences, gardens, and so on) or lack of maintenance that might cause problems in years to come, the warranty rep can point out these conditions to the home owner. Documenting such details can prevent future debate over subsequent damage.
* Arriving with a checklist helps establish the authority of the warranty rep and puts him or her in a better position to deny excessive warranty requests. Warranty reps arrive at appointments that *they initiated* feeling as if they are doing something for the home owner rather than arriving feeling subservient. The difference is subtle, but significant.
* Builder-initiated warranty visits present an opportunity to review key maintenance items and answer questions.
* Communication occurs with home owners who might otherwise be silent. Silent does not equate with satisfied; uncovering some disappointment may allow the company to recapture goodwill that otherwise would have been lost.
* Some home owners may opt not to take advantage of the opportunity for this visit. Even so, they will be favorably impressed with the offer.
* Builders using this approach report less rather than more warranty work. Home owners who are favorably impressed rather than feeling abandoned and angry are more likely to do normal maintenance items, and revenge lists become less common.
* With shorter warranty lists to screen and process, the builder can get all items completed faster.
* This approach benefits staff schedules because it helps to smooth out the peaks and valleys. Staff members generally believe they have more control over their work schedules.
* Typically, fewer home owners will submit year-end warranty requests after the end of their warranty year has passed because they "got busy and forgot."
* When visiting a home under such positive circumstances, the warranty rep is more likely to ask for and get referrals.
* The quality management data that accumulates from a systematic review of all homes can be instructive for design, purchasing, construction, and your trades.
* Perhaps the most significant advantage of this method is that the warranty rep makes the list. Experience shows the validity of the theory that the simple act of writing down an item on an official warranty request form convinces

the home owner that he or she deserves the requested attention. Home owner disappointment or company expense can follow if the item falls within your company guidelines.

A balanced discussion of this topic requires acknowledgment of the potential disadvantages of builder-initiated warranty visits. Note that most of the following items marked with an asterisk (*) also can and do occur with a traditional system that puts responsibility for initiating warranty attention solely on the home owner:

- If your system misses a home owner, ill will can result.
- Warranty reps will invest time arranging and confirming appointments.*
- Warranty reps will sometimes be stood up and need to reschedule an appointment.*
- A few home owners will turn setting up the appointment into a power struggle: "You have to come on Sunday, that's the only day I'm off"—and so on.*
- Dormant issues may resurface. Warranty reps will need to review the file and be aware of and prepared to respond to such controversial subjects.*
- Home owners may disagree with the warranty rep's decision on items the home owners point out.*
- Trade turnover by year end is possible. You may agree to fix legitimate warranty items that the home buyer might otherwise never have reported. You also may issue work orders for these items to new trades who will want to be paid for correcting mistakes made by the previous trade.*
- A home owner might hold off reporting something until one of the routine builder-initiated warranty visits when the item needs immediate attention. The result is a more costly repair.*
- Some home owners will want more frequent warranty attention than two standard visits will provide.*
- You should update your home owner guide to reflect the builder-initiated warranty visit procedure.
- Builder-initiated warranty visits will not entirely stop reports of out-of-warranty items.*
- Dishonest home owners will take advantage of any errors or problems in the service provided through the builder-initiated warranty visits to extend their list of exaggerated complaints.*
- You should update your home owner guide and related paperwork to reflect the procedure for the builder-initiated warranty visit.

If after considering both sides of this discussion you conclude that builder-initiated warranty visits are the best approach for your company, you can implement them in four steps: develop supporting paperwork, establish a method for setting up the appointment, plan a confirmation, and train warranty reps. These four steps are described in the following paragraphs.

Develop Supporting Paperwork

Begin by creating one or more warranty visit checklists. Figure 4.5 offers an example to get you started. You can use the same checklist for the first and the year-end visit, or you may prefer to have two different checklists. Such checklists help to establish the warranty rep as an authority on company warranty guidelines and makes denying excessive requests easier for him or her.

To minimize the "official" tone of the traditional warranty service request form, you can consider a more casual approach such as the Warranty Clipboard form in Figure 4.6. Home owners can make note of items or questions for the warranty rep to respond to during this appointment.

The welcome to warranty letter can be reworded to accommodate the new procedure. You can see an example in Figure 4.7. Once you have this procedure working smoothly, add information about it to your home owner guide. Figure 4.8 offers suggested wording.

Establish a Method for Setting Up the Appointment

Any method that circumvents setting up this appointment prior to closing sacrifices the procedure's greatest benefits. Letters, postcards, or phone calls after move-in fail to work as well. Set up the appointment at the orientation if at all possible. This appointment is your most important transition service.

Some builders set up both of these appointments at the orientation and a few provide a refrigerator magnet on which they document these dates. The same magnet includes space for appropriate phone numbers. (Figure 4.9 shows one possibility. Figure 4.10 shows a proactive end of warranty letter you can send in the tenth or eleventh month.)

Builder-initiated warranty visits at the end of the warranty are just as valuable as the initial 30 to 60 day visits. The expiration of the warranty is a turning point in the builder-buyer relationship—a separation that the builder looks forward to and the home owner resists. Mark this turning point with a ceremony that calls attention to the transition—a sort of graduation ritual that marks the shift in responsibilities in the home owner's mind. Marking the end of the warranty gives you an opportunity to create goodwill, learn which products and materials perform best, ask for feedback on services and design, and obtain referrals. An end of warranty visit can be the most significant part of your end of warranty transition service. Many procedures, such as those that follow, are possible:

- Schedule a team to come in and detail the home one last time. Caulking, filters, batteries, light bulbs (especially those that are tough to reach), and so on might be on the team's list.
- Give the home owner an anniversary gift—a new welcome mat with your company logo modestly displayed is one idea.
- Provide a coupon for a few hours of a handyman service.

A deliberate strategy for managing this rite of passage helps smooth the transition.

FIGURE

4.5   Builder-Initiated Warranty Visit Agenda

[Logo]

Warranty Visit

Home Owner ___Cindy Buyer___   Date ___6/10/20–___ Time ___8:30 am___

Address ___3399 West Port Ave___   Community ___Bailey's Cove___

Home phone ___555-555-1234___   Home site no. ___3-27___

Work phone _____   Plan ___Trinity Bay___

Cell phone _____   E-mail _____

Exterior

☑ Backfill
☑ Drainage
☑ Downspout extensions
☑ Concrete flatwork

Interior

☑ Front door
  • Lock and deadbolt
  • Threshold
  • Weather stripping
  • Doorbell
☑ Back door
  • Lock and deadbolt
  • Threshold
  • Weather stripping
☑ Patio door lock
☐ Garage overhead door
☑ Smoke detectors
☑ Furnace filter
☑ Interior doors
☑ Interior trim
☐ Cabinets
☑ Tile
☑ Caulk
☑ Window operation
☑ Drywall
☑ Floor coverings

☑ Home Owner list? None

Follow-Up Notes

Garage O/H door needs an adjustment —
catches and makes excessive noise

Cabinet to left of refrigerator missing a shelf —
also deliver extra shelf clips to home owner.

Talked w/Mrs. B. regarding the need to main-
tain downspout extension in the down position.
Send follow up letter to emphasize.

Home Owner ___Cindy Buyer___

Home Owner _____

Builder ___Harry Beasley___

FIGURE
4.6 Warranty Clipboard

[Logo]   Warranty Clipboard

Warranty reference		Warranty visits		Office contacts		Emergency contacts	
Home site	3-27	60 day	6/10/20–	Rep	H. Beasley	After hours	555-555-8686
Floor plan	Trinity Bay			Phone	555-555-8686	HVAC	555-555-7766
Closing	4/3/20–	Year end	2/10/20–	Fax	555-555-5555	Plumber	555-555-9999
Expiration	4/3/20–			E-mail	hjb@build.com	Electrician	555-555-5638

Questions

Need booklet for self-cleaning oven.
What do you use to clean jetted tub?

Comments

Poor quality fence!

Warranty Review Items

Fence is poor quality, wood splits, loose
posts, gate is crooked
Garage foor is noisy

Need shelf for cab next to refrigerator

One-Time Items

Drywall

Caulk

Grout

Backfill *Already settling–shouldn't it be filled NOW?*

FIGURE

4.7   Welcome to Warranty: Confirmation of Builder-Initiated Visit

Dear <Home Owner>:

On behalf of [Builder], I'd like to welcome you to < community> once again. We all hope that your move went well and that you are enjoying your new home.

While we believe that we delivered an excellent home to you, we recognize some items in the home may require follow-up work. Our limited warranty spells out the services we provide in this regard.

We want to confirm the appointment we set up with you for a 60-day warranty visit on

_____<date>_____ at _____<time>_____.

This meeting has three purposes.

- We would like to confirm that the home we delivered to you is performing to [Builder] standards. You can review a sample of the checklist we will use in your Home Owner Guide, page <page number>.
- If you have noticed warranty items you believe need attention, we will review them with you and make a repair determination.
- We will answer questions you may have about the operation or care of your home.

We will contact you a few days prior to the appointment to confirm. Meanwhile, if you notice any nonemergency warranty items that need attention, please note them on your *Warranty Clipboard* to mention at our first visit with you. I am enclosing a copy of that form for your convenience; you also have one in your [Builder] Home Owner Guide.

Emergencies are rare, but if one occurs please call our warranty office during normal business hours, Monday through Friday, 8:00 a.m. to 5:00 p.m. In the event of an after-hours emergency, refer to the Emergency Phone List sticker inside your kitchen cabinet.

You can find complete details about our warranty procedures and guidelines in your Home Owner Guide, Section 8, "Caring for Your Home."

Please feel free to call me if you have any questions. I look forward to working with you in the coming months.

Sincerely,

Warranty Administrator
[Builder]

Enclosure

FIGURE

4.8 Home Owner Guide Entry: Builder-Initiated Warranty Visits

Scheduled Warranty Visits

During your Home Owner Orientation, [Builder] sets a tentative appointment with you to revisit your home in approximately 60 days. This follow-up visit has several purposes. During this meeting we will do the following.

☐ Confirm the home we delivered to you is performing to [Builder] standards. You can review a sample of the checklist we will use on page <page number>.

☐ If you have noticed warranty items your believe need attention, we will review them with you and make a repair determination.

☐ We will answer questions you may have about the operation or care of your home.

Confirmation Call

[Builder]'s warranty office will contact you several days prior to the appointment to re-confirm the date and time. At that time we will also ask that you forward your list, if you have noted any items, so that we can do any necessary research and schedule an appropriate amount of time for our visit with you. If you happen to note any additional items between then and your appointment, we will add them to your list as we go through your home with you.

Warranty Clipboard

To make keeping your questions convenient we have included a Warranty Clipboard form at the back of this section. Please use it to jot down items to bring to the attention of your warranty rep.

Year-End Visit

At the end of our meeting with you, we will tentatively schedule a year-end visit 30 to 60 days prior to the expiration of your warranty. A sample copy of the checklist we will use at that appointment appears on page <page number>.

---

* Have a member of upper management conduct an in-depth interview of the home owners to gather feedback about the experience your company provided for the new home owner.
* Conclude with a gift certificate for dinner at a popular restaurant.

Such ceremonial details mark this event in the home owners' minds, and emphasize that the care of their homes is now the responsibility of the home owners. Simultaneously, investing time and attention demonstrates your appreciation of the home owners and your commitment to service.

FIGURE
4.9   Warranty Magnet

```
[Logo]

Warranty Office
Phone _____
Fax _____
E-mail _____

Warranty Appointments
60 Day _____
10 Month _____

After-Hours Emergency Numbers
    HVAC        (555) 555-1234
    Electrician  (555) 555-5678
    Plumber      (555) 555-9999
```

Plan a Confirmation

Several days prior to the appointment, the warranty rep or administrator should contact the home owner to confirm the date and time. Ask whether the home owners have noted any warranty items, and if so, request that they e-mail or fax the list to the warranty office. In this way the warranty rep can arrive better prepared.

Train Warranty Reps

The concept of looking for items to repair may be new to veteran warranty personnel. Plan some training for this task and practice in homes ready to deliver. Role play each part of such visits, from the greeting and introduction of the procedure to discussion of common items likely to be noted and setting expectations about repairs to come, if appropriate. This preparation offers an excellent opportunity to review warranty guidelines and tour model homes to fix the company's promised quality clearly in everyone's mind. Case study discussions to provide your staff with judgment call practice as discussed in Chapter 5 will add to their effectiveness.

Miscellaneous Requests Between Routine Checkpoints

Some home owners may want to report warranty items between your routine checkpoints regardless of how you plan them. An hospitable attitude in these cases allows your system to capture the benefits of both random reporting and

Dear <Home Owner>:

Over ten months have gone by since you closed on your new [Builder] home. We hope you have found your home and the surrounding community to be a pleasant and comfortable place to live.

As you are aware, the Materials and Workmanship portion of your [Builder] Limited Warranty will expire on

_____ <date> _____ .

As we did at 60 days, we ask to visit your home near year end to check on the performance of the home we built for you. We want to confirm the appointment we set with you for a year-end warranty visit on

_____ <date> _____ at _____ <time> _____ .

We will contact you a few days prior to confirm. Meanwhile, if you notice any nonemergency warranty items that need attention, please note them on your *Warranty Clipboard* to mention at our year-end visit with you. I am enclosing a copy of that form for your convenience; you also have one in your [Builder] Home Owner Guide.

You can find complete details about our warranty procedures and guidelines in your Home Owner Guide, Section 8, "Caring for Your Home."

We look forward to working with you. Please call me if you have any questions.

Sincerely,

Warranty Administrator
[Builder]

Enclosure

consolidated lists. As appropriate and in a friendly tone, remind these home owners that they have one or more routine warranty visits coming. If the items are urgent or if the home owner prefers to have the items addressed immediately, proceed cheerfully to schedule an inspection appointment.

Repetitious Requests

Some home owners go to extremes and submit list after list. The goals of customer satisfaction and operational sanity come into conflict when these overlapping lists create confusion and duplication of effort. Further, most warranties state that if a home owner fails to report an item in a timely manner and additional damage results, the builder can deny the claim. Builders need to be cautious about taking too firm a stand on how often items can be reported.

In extreme situations, meeting with the home owner may help. Apologize for failing to make procedures clear and carefully review normal service. These situations can result from delivering an incomplete home or a poorly built home. Resist resenting the home owner who finds and reports large numbers of legitimate items. Get them corrected and work on improving quality control. If home owners submit long lists of items that are within your standards, review how well your company is aligning customer expectations.

Complaints and Questions

Home owners should be able to report complaints about incomplete orientation items, personnel, quality of repair work, or damage that occurred during a repair by phone. You can document these details as well as questions directly into your computer system or on a phone log. Phone logs offer many advantages over small phone message pads because they do the following (Figure 4.11):

- Document emergency items, complaints, and questions.
- Support prompt follow-through by providing a reminder to check back with the home owner.
- Provide sufficient room to describe the situation and to add follow-up notes.
- Note the details of each situation on a separate page, so you can delegate, follow up on, or file.
- Permit periodic review in search of repetitive issues to eliminate in your quality control checks.
- Serve as business records and as such can be introduced in court if needed.

Warranty Reports to Nonwarranty Personnel

You probably seldom hear of a home buyer coming to the warranty office and asking someone there to prepare a purchase agreement. Yet home owners often try to report warranty items to their sales person or the superintendent. While this practice of working outside normal warranty reporting procedures is understandable, it does little to produce the complete warranty documentation that best protects home owners.

At the same time, no one who is concerned about customer opinions should ever utter the words, "It's not my department." This expression amounts to verbal shoulder shrugging. Train nonwarranty personnel how to manage these incidents to prevent situations such as the one described in the sidebar.

FIGURE

4.11 Phone Log

[Logo]   Phone Log

Name ___Cindy Buyer___   Date _7/15/20–_

Address ___3399 West Port Ave___   Community _Bailey's Cove_

Home phone ___555-555-1234___   Home site no. _3-27_

Work phone ___NA___   Plan _Trinity Bay_

E-Mail ___NA___   Closing date _4/3/20–_

Message   **Response**

Message	Response
Need to reschedule inpsection of basement water.	Paged Harry B., no problem
problem; has to go to doctor. Can reschedule for	2:00 tomorrow works.
tomorrow, 2:00 pm	

Follow-up notes

Confrmed that 7/16 appt still works for Mrs. B. Harry on the way.

By _____Alicia A._____

A home owner described his drainage concerns to his salesperson who said, "Warranty will address that for you." Two weeks later the same home owner crossed paths with the superintendent and again mentioned his drainage concerns. The superintendent also reassured the home owner that this was an issue warranty would address. Thinking that sales and the superintendent had both alerted warranty, the home owner did not contact the warranty office. Two more weeks went by and the home owner became enraged. Meanwhile, the warranty office had no clue the home owner had a problem. Neither sales nor the superintendent passed the complaint along to warranty.

Provide frontline staff with warranty service request forms or computer access to online warranty requests. Instruct them to guide the home owner in reporting the warranty items or even to complete the request personally if the home owner is upset.

"The fastest way for us to get this issue addressed is to get it into the warranty system. Here is a service request form. Let's fill it out and fax it to the warranty office." Another reasonable approach is "The person who can help us with that is Susan. Let me get her on the phone for you." This response is much different from, "You'll have to call Susan about that."

Emergencies

Your definition of emergency items should appear in your home owner guide with a clear explanation of how to report emergencies. During normal business hours, builders typically take reports of emergencies by phone and address them immediately. Document emergencies directly into the computer or on a phone log. Builders typically manage emergency service outside of normal business hours by one of three methods. Whichever method you use, test your emergency system from time to time.

Emergency Phone List

A common approach is to provide home owners with a list of emergency phone numbers for critical trades (HVAC, plumber, and electrician). Document the after-hours obligation in each trade's contract. Include a recommendation in your home owner guide that home owners who report an emergency directly to a trade outside of your normal business hours also notify you. This procedure allows you to follow up and keep their homes' warranty histories complete.

On-Call Duty

Another option is to establish on-call rotation among warranty personnel. In small organizations, construction personnel are often included in this responsibility. Typically, on-call duty is assigned for a week at a time, and each individual in turn carries the company pager and a list of the trades available for emergency service.

Outsource

Some builders prefer that their home owners reach a live voice and therefore use an answering service for after-hours service. If you take this approach, be certain you understand fees, what services are available, how much scheduling flexibility you have, staff stability, and the reliability of the answering service company's technology. To prevent confusion, assign one person to manage the relationship with the answering service firm.

Service Request Log

Once you have established your warranty structure, you need systems for managing the details that flow through it accurately and efficiently. Either by computer or on paper, track receipt and disposition of miscellaneous warranty service requests with a simple log that shows date, home owner, number of items, and to whom the list was assigned. When Ms., Mr., or Mrs. Jones calls asking "Did you get the list I sent last Friday?" anyone can check the log and determine that it arrived on Monday and which warranty rep will address it. This log also enables you to track how many service requests are coming in. Over time a seasonal pattern emerges, and trends become easy to spot if a sudden increase in service requests occurs.

Put It in Writing!

Thanks to the Internet, the age-old debate over whether items are reported in writing has mostly fallen by the wayside. Submitting items via an e-mail to the warranty desk is fine with most home owners. Even so, no discussion of new home warranty would be complete without touching on this topic.

Historically, home owners have tried to report their warranty requests by phone. Home owners often followed this trend not because that method is more convenient—after all, if the home owners had more than three or four items, you can imagine that they had written them down so making the list wasn't the issue. Instead, they use the phone because they believe that talking to a live person will produce results. When effective warranty processes produce a timely response, most home owners don't object to submitting items in writing.

In spite of the fact that limited warranty documents (and increasingly state statutes) require that home owners submit items in writing, some builders fail to enforce this requirement. This situation is another one in which what appears on the surface to be better service turns out to be just the opposite. Service quality sometimes collapses under the volume of calls.

Small-volume builders may be tempted to take items by phone because few homes are involved. By establishing the habit of getting items in writing, you will avoid setting a problem-causing precedent as your company grows. Mid- to large-size companies can fall into the trap of taking items by phone as well. One builder who accepts items by phone and has one warranty administrator recently experienced 250 calls in one week, 63 of them in one

Regardless of company size, if builders exercise self-discipline throughout the home building process and document every detail, home owners usually follow the builder's example during construction and continue this practice during their warranty period.

day. At that volume, nothing else gets done. Courtesy and effectiveness suffer with mental health not far behind.

In another situation, an administrator was on the phone nearly 40 minutes working with a talkative home owner who had 14 items to report. Meanwhile 7 messages accumulated including one about a serious plumbing leak. Without adequate staff, your company personnel, trade contractors, and home owners seethe in frustration at the inefficiencies of a system that produces more busyness than results.

After considering these potential difficulties, if you still want to take reports of items by phone, have enough staff to answer calls promptly. (Keep in mind that adding staff adds overhead and that price increases may result.) Staff should document each item reported and e-mail, fax, or mail a hard copy of the list to the customer for his or her records. Prevent this issue from turning into a conflict in the first place by volunteering information about service procedures early. As with so many other issues, *volunteering information* means covering the information in the home owner guide and talking to home owners early in the process about the benefits of the builder's procedures. Tell the home buyers that the requirement to provide written lists of items to be inspected for possible correction or repair protects both home owner and builder as follows:

- If a recurring problem persists after the warranty has expired, a solid paper trail provides the best chance for additional service for the customer.
- Home owners describe the items in their own words, without any editing or abbreviation by a staff person rushing to answer the other line.
- Staff time is invested in producing results rather than serving as scribes.

When home owners call to report a list of items, unless an emergency exists, say "Yes, these items are the kind that you would list on a warranty service request. You should have two of them in your home owner guide, or I can fax one to you. You also can use our Web site service request:" This response allows the home owner to save face and still agree to follow the procedure.

You should accept reports sent by fax or e-mail because they achieve the same objectives as a mailed list with greater speed. The goal is to permit the warranty staff to work on producing repairs.

Training

The suggestions listed here will take from 10 minutes to about an hour. Include short training exercises as part of routine staff meetings and plan longer sessions on a regular schedule. Training aims to ensure that staff members follow procedures and know how to manage exceptions as follows:

- To get your team working smoothly and consistently, begin by reviewing all letters and forms used to gather reports of warranty items. Start with blank forms and examine correctly completed versions as well as common errors to avoid.

⬡ Discuss examples of the value of gathering factual background information before meeting home owners for the first time.

⬡ Log into your online system and "report" some warranty items to see how the system works for home owners.

⬡ If you opt to follow the builder-initiated warranty visit procedure, remember to train your field personnel in this new attitude and method as described earlier in this chapter.

⬡ Work with salespeople and superintendents to ensure their comfort level with managing home owners who want to share warranty requests with them.

⬡ Review your definition of the term *emergency;* then discuss common sense responses to urgent items that fail to qualify as emergencies but need prompt action, such as no hot water or a deadbolt that fails to lock.

⬡ Role play typical conversations such as the method of
  – introducing the warranty rep to the home owners
  – asking a caller to report warranty items in writing with emphasis on the benefits of this approach to the home owner
  – setting up an end of warranty inspection appointment and so on

Repeat each assignment with different players so staff can hear variations. Ask them for suggestions on new scenarios to practice as well. Collect examples of well-handled requests for warranty service and examples of those that could have used more thought. With staff member's and customer's names deleted, you can incorporate these examples into additional training or review sessions.

Decisions

CHAPTER 5

I n the warranty process, builders strive to balance specific legal obligations, an inevitably tight budget, and the home owners' expectations. The most clear-cut situations require timely and courteous responses, involving accurate communication of many details. However, much day-to-day warranty work is made up of unique situations, fuzzy boundaries, personal opinions, paradoxes, contradictions, exceptions, judgment calls, and even an occasional honest mistake. The analysis step of warranty processing involves deciding whether an item should be fixed and, if so, who should do the fixing. Much of the time, an inspection is needed to determine appropriate action.

Beware of "Desk Decisions"

Some builders make the mistake of having an administrator receive warranty requests and decide at his or her desk whether items listed are covered by warranty. Work orders for approved items are issued to trades without screening. These desk decisions often lead to providing service that should have been denied, denying service that should have been provided, sending the wrong trade, or fixing cosmetic symptoms and missing a serious underlying structural cause. Home owners may not describe an item accurately to begin with. They may omit pertinent details, exaggerate the problem, or fail to understand what normal home performance standards are.

In addition to providing the home owner with more appropriate responses, by screening items with an inspection, you show respect for your trades' time, show care and interest in the home, and gain valuable insights into the quality of your company's product. Especially when the answer to a customer's warranty service request is no, the courtesy of a personal visit can help your company retain the home owner's goodwill.

Control Your Part

In a world with instant messaging, overnight delivery, and microwave ovens, no wonder home owners expect prompt warranty service. In that same world, builders create homes with the support of 35 to 50 individual trade contractors for whom production produces income while repairs do not. No wonder delivering prompt warranty service is a challenge.

Admittedly, some timing elements are outside of your control. The home owners' and the trades' schedules must be coordinated. If parts need to be ordered, the process becomes even more complicated. However, controlling the steps that you can makes a significant positive impact.

If you are using the builder-initiated warranty visit procedure described in Chapter 4, the first appointment (and perhaps even a year-end appointment) is already tentatively set. If a home owner has submitted a list outside of your standard builder-initiated checkpoints, acknowledge receipt of this request as soon as possible—ideally within four hours.

Strive to acknowledge the request and set the inspection appointment in one contact. Some companies acknowledge the request and tell the home owner it will be forwarded to a warranty field rep, but this process can add many days to the overall response time. Home owners understandably wonder why they can't report items directly to the rep and avoid this extra step.

However your company handles this detail, track the time from receiving the request to setting an inspection appointment. Set your target as a number of hours—not days. Be persistent. Leaving a message asking that the home owners call back to arrange an inspection is fine, but if you do not hear from them within two business days, contact them again. Note the date and time of each attempt.

Although the time between your contact and the inspection appointment is influenced by the home owner's schedule, you should be able to offer an inspection appointment within five business days. If the best you can offer is an appointment three weeks in the future, check your time-management habits first. Next look for other explanations, such as a temporary seasonal overload or some personnel on vacation. If the condition is chronic—lasting more than two months with no end in sight—examine your company's staffing level. Managers who resist investing in an adequate staff should consider the financial impact of losing sales because of slow warranty service.

Scheduling Inspection Appointments

When following the builder-initiated procedure, your product line and the details of your agenda will determine the minimum amount of time to allow per home. Keep in mind that this visit is intended to be proactive; you are meeting with the home owner to review the home on the home owner's behalf. Start with a minimum of 30 minutes and adjust up or down from there as experience dictates.

When the home owner initiates contact by submitting a list, the time needed naturally will depend on the number and nature of items listed as well as the personality of the home owner. You may need from 10 minutes to several hours for an inspection appointment. The variables involved can make scheduling inspections difficult, but the suggestions that follow can make scheduling easier:

- Allow sufficient time to review and discuss each item, particularly when the home owner has submitted a list of items you suspect to be home maintenance.
- Begin with an estimate of two to three minutes per item (10 items, allow 20 to 30 minutes) until you get to know the home owners and become familiar with floor plans and typical items.
- Allow a buffer between appointments, particularly if you will be driving to another community. If the meetings are all in the same community, 10 minutes is usually sufficient.
- If your company builds in several locations, consider scheduling visits to certain communities on certain days of the week. Think of this idea as a guideline, not a hard-and-fast rule.
- Especially if this warranty appointment is the first one with a new home owner, explain that you are coming to determine what attention is needed and that the repairs will occur later. If you carry tools and perform some repairs, explain this situation and the subsequent role of trade contractors.
- Immediately record each appointment in your appointment book or electronic schedule.
- Call one or two days before the appointment to reconfirm the time.

Preparing for Inspections

To prepare for an inspection, review the list and conduct obvious research. Figure 5.1 lists sources of information you might consult to arrive at fair decisions. Avoid drawing conclusions prior to seeing the items. Home owners often describe items inaccurately, so decisions made before seeing the items may be incorrect.

Your home owner guide is an essential tool to carry on warranty appointments. In addition, by keeping your materials readily available, you will be more productive and less stressed. Designate a place in your office for items you need for inspections:

- the warranty visit agenda for a builder-initiated appointment or the list the home owner submitted
- pen and inspection form, handheld computer, or electronic notebook
- digital camera
- business cards
- tape measure
- your community binder or an electronic community file (see Figure 5.2 for suggestions on contents)
- a four-foot level (in your vehicle in case you need it)

Make your approach to doing inspections systematic and consistent. Prior to the appointment, review the orientation list and any previous warranty work orders, and consider whether checking with the salesperson or superintendent would be helpful.

FIGURE

5.1 Information Sources for Inspections

- purchase agreement and addenda
- selection sheets and change orders
- quality management inspections
- orientation list
- warranty document and guidelines
- previous warranty requests
- previous inspection reports
- phone logs
- correspondence
- photos or diagrams
- product specifications
- manufacturer installation instructions
- construction personnel
- sales staff
- trade contractors
- manufacturer reps
- building department inspectors
- code manuals
- engineer
- attorney

FIGURE

5.2 Community Binder Contents

Ever have a hard time locating lot number 41 in that community that just opened a few months ago? Give yourself (or your front line personnel) a stress relieving gift of readily available information, gathered and well-organized in community binders. Standardize the contents as much as possible to make creating binders for new communities easier.

- Begin with a map clearly showing the location of the community.
- Add a map of the streets *within* the community. Note significant landmarks on this map if the community is large.
- Add the 8.5 × 11–inch copies of floor plans from the sales office.
- Include the list of features.
- Include the list of available selections.
- Provide a directory of home owners: names, addresses, phone numbers, floor plans, and closing dates. Include phonetic pronunciation for names where appropriate.
- List the names and contact information for on-site personnel—office, fax, e-mail, pager, and so on.
- Repeat for the trades and associates (such as the management office of the home owners association) involved in that community.
- Emergency numbers such as local authorities, hospitals, locksmith, water extraction company, and so on may seldom be used, but they are invaluable when needed.
- Over time, you might add reference photos to this binder.

For some companies, one binder can hold information about all communities; for others, several may be needed. The key is to have ready access to these details when you are parked in front of any home in any community. Not only can you find what you need more conveniently, but when the trades come to you needing such details, your ability to answer their questions can lead to faster responses for home buyers.

Performing Inspections

Be on time. If the home owner is late, wait a minimum of 10 to 20 minutes. If you're meeting the home owner for the first time, introduce yourself, shake hands, and provide a business card. Chat for a moment if the home owner's personality makes that appropriate—10 percent of the appointment time works well. Finally, introduce the inspection process.

Review the items listed in as logical an order as possible. Ideally you begin on the outside, at the street. Review the exterior of the home, enter through the front door and end in the kitchen. If your company's frame stage-tours and home owner orientations follow this itinerary, the home owner is accustomed to it. Equally important, you will avoid overlooking items.

Frequently you can make inspection notes directly on the warranty request form. If this is impractical, an inspection form similar to Figure 5.3 is an alternative. Complete the top portion prior to the appointment. List items in the order inspected and note your decisions legibly. As an electronic alternative, Figure 5.4 shows notes taken on an electronic notebook. The computer converts the hand-written words to type so you can forward them to the appropriate trade. This procedure reduces response times and home owners are impressed with this professional approach.

In explaining your decisions to the home owner, you have two choices. One method is to explain what action you will take as you inspect each item. A second approach is to look at all the items before committing to specific repairs. Seeing all the items may influence your answers. For instance, after you view the drywall crack in the secondary bedroom, you know that the painter will be in the home, and you may as well have him touch up that spot in the entry hall you would otherwise have turned down. Match your approach to the situation and the home owner's personality.

Judgment Calls

Most builders fully intend to do the "right thing" for their home owners. Sometimes, however, knowing what that "right thing" is can be difficult. As discussed in Chapter 3, warranty guidelines are only a starting point. Events can occur in a new home that builders should correct even though they are not addressed in the warranty guidelines. As much as builders would like to have standards that provide answers to every situation, the reality is that common sense and good judgment are essential. You may be called upon to make judgment calls at any of three levels—physical, circumstantial, or image.

Physical Judgment: The Documents and the House

Physical judgments are usually the easiest. The terms of the purchase agreement or the limited warranty often provide clear answers. Comparing physical conditions to measurable standards creates confidence—a quarter of an inch last week

FIGURE
5.3   Warranty Inspection

Warranty Inspection

Home Owner __Melinda Matthews__   Date __8/13/—__   Time __9 am__

Address __628 West Port Lane__   Community __Bailey's Cove__

Home phone __555-407-9876__   Home site no. __2-11__   Plan __Bay Pk__

Work phone __NA__   Plan __6/12/20—__

Cell phone __NA__   E-mail __MMatthews@hotmail.com__

Item	Action	Work order number
Crack in drive	No action; advised home owner to caulk	None
Dead shrub	Gave home owner discount coupon for nursery; shrub appears to have insect problem	None
Attic access panel split	Dale's Drywall should replace; attach PO This could have occurred during construction and been overlooked; I've seen this prior in other homes	7499

Notes

I was unable to complete the inspection due to Mrs. Matthews disagreement with Roger from Carl's Cabinets—see incident report.

Home Owner _____   Date _____

Home Owner _____   Date _____

Builder __Harry Beasley__   Date __8/13/20—__

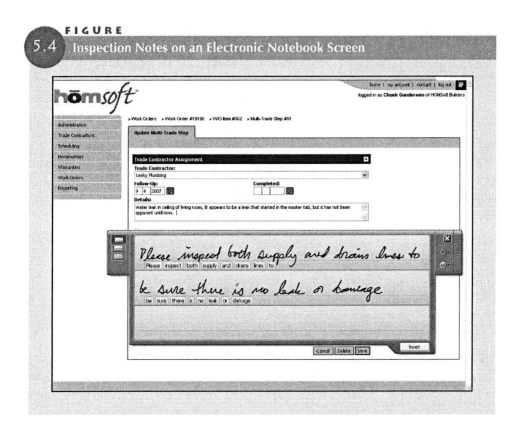

5.4 Inspection Notes on an Electronic Notebook Screen

is the same as a quarter of an inch this week. Corrections provided at no cost by the trades are usually approved at the physical judgment level.

When a review of the paperwork or the home results in saying yes to warranty service requests, provide the requested repair, replacement, or correction. If you decide that the request is not justified, explain this decision to the home owner. Use your home owner guide to show the home owner the steps to take for the maintenance item. Rather than say "That's not covered by your warranty," or "We don't fix that; you have to do it," you can say, "This is a maintenance item," or "Let's check the manual and see what we're supposed to do with this. . . . Okay, here it is. This is a maintenance item. I'll be glad to answer any questions you have about taking care of it."

Circumstantial Judgment: The Home Owner's Side of the Story

Even with the guides, manuals, and established procedures, warranty service cannot be managed from a book of printed standards regardless of the author, number of pages, or level of detail. Circumstantial decisions are inevitable. They begin when a home owner says "Yes, but you don't know what led up to this situation." followed by the home owner's side of the story. At this point, you may face a circumstantial judgment, where standards are much less well defined.

Listen carefully—customers often have quite valid points. Sift the facts out of the stories home owners tell you. Their perspectives usually include bits of information unnecessary to your decision. At the same time, some relevant points will likely be omitted from the home owner's description of events, and you may need to conduct some research to fill in those blanks.

Avoid decisions based on your personal opinion of the home buyer. Asking "What was it in our information that made you expect . . . ?" is often helpful. If the home owner can point to something in the purchase documents, the decision may be obvious (and revising the documents may be appropriate). Back up careful listening with research. Check original documents, look at the model home, and talk to other personnel. Then use common sense to arrive at a fair answer.

Circumstantial judgments—sifting facts from rationalizations to arrive at that "right" answer—are usually tougher than physical judgments. Yet even circumstantial judgment does not generate the consternation of the third level, image judgment.

Image Judgment: Reputation at Risk

This third level of judgment is the most complex and the toughest to master. Assume that after you inspect the home, review the documents, and listen carefully to the home owner's explanation of the circumstances, you conclude the correct answer to the request is no. However, the home owner's personality makes it clear that no is the wrong answer. If you say no in making this image-level judgment, one or more of several events, all unpleasant, can follow. For instance, low survey scores, negative word-of-mouth, complaints to the salesperson, a sign in the yard, media attention, arbitration, or litigation—to name a few.

At this point, you may be tempted to provide the requested attention, but before giving in, consider the potential repercussions. Your home owner learns which threat produces the desired result and may repeat it or share the technique with neighbors. Going beyond planned services costs money, and home owners with justified requests wait for the attention they deserve while the intimidator gets extra service. Meanwhile, frontline personnel develop cynical attitudes—such employees are often said to be "burned out." The result is costly turnover and disruption of service and productivity. No simple formula exists for making these tough decisions, but the hints listed below may help:

- Discuss the issues with others in your company. Another perspective or additional background information may help.
- Review the history of the relationship. Company errors may emerge, or conversely, you may identify a pattern of the home owner continually pushing the boundaries.
- Consider the status of the relationship. Did the home owner just close on the home, or is it nearly out of warranty? The time remaining in the warranty may influence your final decision.

- Consider the larger business context. What is the nature of your relationship with nearby home owners? The home owners association? The jurisdiction? Media? Court system?
- Finally, estimate the financial costs in administrative time, repairs, and the impact on work schedules.

After you resolve the situation, review company performance and look for improvement opportunities to prevent similar future problems.

Precedents

Warranty personnel often shy away from setting precedents because they hope to operate entirely within established parameters. This hope is not only unrealistic; it also can be unfair to your home owners (and your reputation).

Precedents are not intrinsically negative. The question is, are you setting a precedent you will be comfortable applying in the future? Would you be willing to provide the same attention to other home owners who had the same conditions in their homes? Consider the following two examples.

In the first situation, during his orientation a home buyer pointed out four footprints in the concrete walk from his drive to the front porch and asked that something be done about them. The builder rep explained with a straight face that because no standard for footprints in concrete exists the company did not have to do anything. The customer appropriately carried the issue up the management ladder. The walk was replaced. This replacement was appropriate and should have been done as part of preparing the home for delivery in the first place. The precedent was a reasonable one to set. The same correction should be provided to future customers in the same situation.

In a second situation, a home owner had convinced himself that he deserved a more expensive type window than those shown in the model and promised in the specifications. His persistent and often threatening requests developed into an ongoing stress for the warranty manager. To resolve the situation, the warranty manager agreed to split the cost of whole-house window replacement with the home owner on condition that the home owner not reveal the agreement to neighbors. Within a week of the work being completed the warranty manager started receiving phone calls from other neighboring home owners asking for the same arrangements. This window replacement is an excellent example of a precedent that should not even have been considered.

Warranty Decision Fine Points

Veterans of the home building industry will be quick to tell you not to say, "Now I've heard it all" because this statement guarantees something even more astounding will occur shortly. Anticipating every problematic situation that could ever occur with home owners is inconceivable, but being aware of lessons others have learned the hard way is extremely helpful.

Avoid sending your third-party experts to inspect items alone. A warranty rep should attend such meetings to provide information and document outcomes. You only need to have one conversation with a home owner who says, "The carpet rep said the carpet needs to be replaced," when the written report from that same representative says, "The home owner needs to vacuum more frequently," to understand why.

Builder's Third-Party Experts

In some cases, you may want the expertise of a trade contractor, manufacturer's rep, testing lab, or building department official to help you reach decisions or convince home owners that your position is correct. When this need is obvious, arrange to have that third person attend the inspection appointment with you.

Building Department Passed It

All warranty staff need to be aware that an inspection by the municipality and issuance of a certificate of occupancy (CO) guarantees absolutely nothing about the correctness of the home. That a municipality could miss something during an inspection is *not* hard to believe; it happens all the time. Using this fact as a defense is inexcusable, both ethically and legally. This statement effectively nullifies any goodwill that could possibly come from providing repair attention. The item in question deserves a fresh investigation instead of a knee-jerk response.

Cosmetic Damage

Nearly every builder orientation form includes a statement disclaiming the builder's responsibility to repair cosmetic surface damage noted after the home owners move in. Although the theory is simple, the reality is much more complicated. At the orientation, typically three or more adults tour a clean, empty home in broad daylight, with the purpose—among others—of confirming the good condition of cosmetic surfaces Although the likelihood of overlooking significant surface damage is low, it can happen. More common is cosmetic damage that appears after home owners move in without their having caused it. Minimally, you owe the home owners the courtesy of an inspection.

Consider this example. If the electrician drops a tool into the tub during construction and is relieved to see no damage had occurred, he may leave behind a crack that is invisible to the eye. After move in, when the home owner uses the tub, the weight and temperature of the bath water cause the chip to appear. Damage occurred after move in, but the home owner did not cause it.

Establishing "back door standards" can help in such situations. These guidelines are not published to home owners, but provide boundaries for discretionary decisions. You might develop yours based on a combination of a number of days, an amount of money, or specific repairs. For instance, back door standards might provide the latitude to take the following actions:

* Repair any cosmetic damage the home owners discover during the first 5 to 30 days after closing.

- Spend up to $200 for cosmetic repairs at any time. (Adjust the amount to suit your product's price point.)
- Replace any window that gets a stress crack.
- Provide a surface repair or patch to a countertop, carpet, resilient floor, tub, or sink.

Note that your cosmetic repair position is different if the damage occurred as a result of a warranty-related repair rather than simply being discovered after move-in. For instance, if the plumber was fixing a leak and in the process scratched the countertop, the customer's expectation of a replacement is more justified than if the scratch has an unknown origin and was discovered after the home owner lived in the home for six weeks. In the former situation, everyone agrees the customer did not cause the damage. The plumber is responsible for protecting the work area and should be held accountable for the needed repair. While a cosmetic repair may be possible, it had better be good.

Listing Agent List

Listing agents sometimes become so excited about their work they feel everyone else should treat their needs with the same urgency as they do. ("I need you to replace the driveway, plant two more trees, and have the carpet cleaned by Thursday so I can show the property.") Show courtesy and respect as you explain the normal reporting procedures by saying, "I'll be happy to help you. Here's what we need to do. . . ." Inspect and evaluate the items based on normal company procedures and guidelines. Maintenance is the home owners' responsibility even if they are selling their home.

Orientation Items Incomplete or Unacceptable

When a home owner calls the warranty office—often frustrated or angry—to complain that "Nothing's been done" about an item from the original orientation list, avoid jumping to conclusions. Home owners should not be asked to document the list again—these items were already documented at orientation. Get the original list and go over it with the home owner or schedule a time to see for yourself. After confirming which items are incomplete, communicate immediately with company personnel to get a commitment on when the work will be performed. Follow up with the home owner to confirm that these items are addressed as promised. If this complaint is common at your company, provide objective reports and quotes from home owners to those who have the authority to correct the causes. (Yelling at them usually won't help.)

Out of Warranty

The expiration of the limited warranty does not free the builder from legal and ethical obligations. Follow the same steps that apply to warranty requests submitted during the warranty period.

> Many circumstances justify providing repairs outside the warranty period.

Grace Period. These periods vary from 10 to 30 days after the expiration of the warranty. A grace period allows a home owner who notices a warranty item on the last day of the warranty a reasonable time period to report it.

Code Items. Builders must comply with the codes that were applicable at the time of construction regardless of the status of the warranty.

Contract Items. Builders must fulfill the contract. If the buyer ordered the optional shelves over the laundry and the builder forgot to install them, they must be installed (or the money refunded) even if the home owner does not notice the omission until the warranty has expired.

Latent Defect. A defect that could not be discovered through normal inspection but existed from the beginning is a latent defect. For example, incorrectly installed valley flashing allowed a slow roof leak. Damage appeared when the home was three years old and a living room window sill began to rot from the moisture. The builder correctly provided repairs.

Written Notice. If the home owner reported an item in writing during the warranty period, but the company failed to respond or cannot prove it responded, the builder must now respond. This example shows the usefulness of bringing closure to each item reported—either with a work order or a documented denial.

Recurring Items. If the same problem was repaired twice or more during the warranty, the failure to repair it satisfactorily might subject the builder to a breach of warranty claim. This concept does not apply to routine repetitive maintenance tasks such as caulking. But if the air conditioner compressor misbehaves the same way it did during the warranty period, the builder's obligation continues.

Manufacturer Covered Claims. Consumer product warranties often provide protection for the home owner beyond the builder's limited warranty coverage. The home owner may require assistance from the builder in these matters, especially because the builder will have more clout with the manufacturer. Builders should provide assistance not only to help the home owner, but to learn how well the manufacturer stands behind its products.

Personality Conflict

No one expects that every home owner will be your favorite. When you work with a large number of people, personality conflicts are possible. Cultivate unflappable courtesy and an ability to work with a variety of personalities and communication styles. When you encounter friction, identify the differences between your two personalities. Discover what you can emphasize more to help you get along with the home owner. For example, engineers are notoriously detail oriented. Make a special effort to respond with precision and thoroughness. In extreme cases, consider whether another staff person can take the lead in working with a home owner whose personality clashes with yours.

Second Owners

If you are contacted by a second (or third or fourth) owner with warranty claims, respond courteously and promptly with the following steps:

- Offer to meet with the new home owner; set an inspection appointment.
- Prior to the appointment, review the warranty file including the orientation list.
- Confirm the status of any pending work orders.
- Pay particular attention to items previously denied and the explanations given.
- Gather copies of appropriate documents to provide to the new home owners. These documents would minimally include the applicable home owner guide.
- Be gracious in your meeting. Introduce yourself, provide a business card, and welcome the new home owners to the community.
- Listen to what the new home owners have to say and inspect the items they reported.
- Provide appropriate repairs.
- Explain the repair process, if applicable, and estimate the time frames involved.

Stand Behind Your Product. Your company has no obligation to provide maintenance repairs promised by previous home owners or any real estate agent involved in the transaction. However, the company should still stand behind its product as it would for the first home owner. Some companies state in their limited warranty that coverage terminates upon sale of the home. Especially if serious items are involved, this attitude is unlikely to hold up well in a court of law and will certainly not hold up well in the court of public opinion. To home owners, this position appears to be a ploy used to sidestep legitimate responsibilities on a contrived technicality.

Seller's Mistaken Promises. If you decide that the company has no obligation while the new home owner insists the first owner or a real estate agent promised action, document your answer, and suggest that the new home owner consider contacting the previous home owner or real estate agent involved. Take the approach that because the sellers "inadvertently created a misunderstanding," they may feel an obligation to help make things right.

Third-Party Reports of Items

Items discovered by the home owner's relatives, friends, real estate agent, or a hired inspector are subject to the same analysis and guidelines as items the home owners discovered on their own. Your company does not have a second edition of the home owner guide with higher standards, especially for third parties. Give respectful explanations that show your procedures and standards are subject to common sense but not negotiation. If home owners forward an inspection report, graciously accept it for review and inspection. Refusing the inspection report aggravates the home owner who can then resubmit the same list over their

Because the items included in a floor plan may change in response to market conditions, builders are wise to ensure that the feature lists for each floor plan include the implementation date. Warranty personnel should have copies of this literature readily available to ensure every customer receives all the promised features.

own signatures anyway. Your concern is not the source of the list, but the condition of each item on the list.

Why Don't We Have . . . ?

Faced with this remark, you should avoid assuming the home owners are mistaken. Get on their side and say, "Let's start by going through your file." Go to the original sources to investigate—avoid trusting memory. Begin with the contract, selection sheets, change orders, model home, and company or trade personnel—any source that can help clear up the confusion. Show the home owners you have checked carefully to confirm they received everything they paid for. If the item was not part of their contract, provide information to help the home owners acquire it now, if possible.

No Inspection Needed

When the list is short and you have no questions about the appropriateness of sending a service technician to perform repairs, the inspection step can be omitted and work orders immediately issued. For instance, a home owner who has shown herself to be well-informed reports that an outlet has no electricity. She further remarks that she has tried several appliances in the outlet. She also says she has checked the ground fault circuit interruption control and the main breaker for a possible solution with no results. Contact the home owners to let them know you have issued a work order, and service is on the way.

Although not every warranty request should be filled, each one deserves objective consideration. Approach warranty decisions with good information and a clear system. Having addressed that challenge, you are ready to tackle the repair process.

Set Up Repair Expectations

Conclude the inspection by explaining what will happen next. Describe the repair process and set a time for an update on unanswered questions. Address details such as the need for the home owner to remove delicate items from the work area, restrain a family pet in a safe place while workers are going in and out of the home, and so on. Covering these details as part of your inspection visit improves the likelihood of a smooth repair appointment.

Training Activities

Training for the personnel who conduct inspections should be planned and ongoing. Many worthwhile activities can be included in a comprehensive plan.

Documents

Warranty personnel training begins with understanding the sales contract and warranty document as described in Chapter 2. Now add to that literature a review of your company's trade contractor agreement and individual scopes of work. Few can memorize all the details covered; your goal is to create awareness and ensure that warranty staff members know where to find answers when questions arise.

Every warranty department meeting should include a review of a section of the home owner guide. Spending just five to ten minutes reviewing one or two topics can keep information fresh in everyone's mind.

Walking Models and Homes Under Construction

Familiarity with model homes is another essential component of warranty training. Walking the models monthly or at least quarterly and studying their details provides a clear view of the promise made to the home buyers and a point of reference on quality. Similarly, walking homes at various stages of construction and asking questions of superintendents and the trades provides a wealth of technical knowledge that helps the staff explain details to home owners.

Observations

Allow time for newcomers to observe veteran staff members performing inspections. Then reverse the process and have the veterans observe the newcomer conducting an inspection. As they discuss the details, both will gain insights.

Digital Photo Classes

By combining digital photos and slide presentations, you can organize short training sessions that focus on one particular warranty item: for example, the function of window components, carpet installation, or cabinet adjustments. The visual nature of these classes makes a strong impression on the entire team. At the same time, the classes achieve efficiency and improve retention of the information. Trades and suppliers may offer prepackaged materials on their products. Over time, develop a library you can draw from on a planned rotation.

Forms and Technology

Review how to complete forms, what standard letters are available, how to develop unique letters for special situations, and how to utilize communication and computer technology to their fullest benefits. This activity can be another routine part of staff meetings or the information can be developed into a longer, more formal training class.

Customer Communication Skills

While hiring people with the right skill sets and personalities for warranty work is essential to long-term success, everyone can improve his or her communication skills. Hardly anything works as well as practice in improving communication.

FIGURE

5.5 Role Play Assignments

Home owners complain about the items in the following list. How would you respond? Ensure that staff members who conduct inspections know the applicable standard or guideline, the sources for more information, the repair provided, what exceptions are possible (and the circumstances that justify them), and how to describe your company response to the home owner who complains about these items:

1. The floor squeaks.
2. Home owner hears noise in the ductwork.
3. Concrete flatwork is cracked.
4. Foundation is cracked.
5. Ceiling sheetrock has ridges or waves.
6. Wood grain or stain doesn't match.
7. Walls are bowed.
8. A window or door is out of square.
9. Driveways are spalling.
10. Nail holes are sunk too deep in siding.
11. Shower floors have some give to them.
12. Home owner wants heat run into the garage.
13. The property has grading/drainage issues, standing water.
14. The glass in French doors sometimes gets scratched as it is being cleaned before delivery.
15. Home owner believes insulation is inadequate.
16. Master bath and master walk-in closet are too cold.
17. Home owner believes that odors in the home are out of the ordinary.
18. Overspray or mortar appears on shingles on roof.
19. Exterior trim does not have enough paint on it.
20. Hardwood floors cup with no apparent source of moisture.

The role-play exercises listed in Figure 5.5 focus on customer interactions that are likely to occur during warranty inspections. Add other situations you have encountered to this list. Again, this activity can be a short review at staff meetings, or you can address it in longer sessions scheduled as more formal training. More material on this aspect of warranty work can be found in the book *Meetings with Clients* (by Carol Smith and published by BuilderBooks).

Warranty Repairs

H ome owners prefer that all of the components in their new home perform properly. Their distant second choice is to have prompt, courteous, and effective warranty repairs. Translation: They want on-time, high-quality work, including clean up of the work area and excluding collateral damage. In other words, they want *results*.

Work Orders

Once you approve an item for repair, the next step is issuing a work order (Figure 6.1). When builders notify the trades of warranty items by phone, the messages have an unfortunate habit of being lost, forgotten, or misunderstood. Adequate tracking is nearly impossible, and builders sacrifice the protection of having accurate documentation in the home owner's warranty file. Documented work orders eliminate these concerns.

Work orders also provide service people with essential repair details. In addition to the home owners' names, addresses, and phone numbers, work orders document color, style, size, and other relevant repair information. Easy access to information about selections and the trades who worked on a home is invaluable to warranty efficiency. Add lifestyle details to further expedite the work, such as "Home owner available on Tuesdays." If your company expects to pay for the needed work, make that commitment clear on the work order.

Hard Copy Work Orders

By printing work orders on three- or four-part, no-carbon-required (NCR) forms that are color-coded, you can send one or two copies to trade contractors (for instance, white and blue), one to the home owner (pink), and retain one for manual tracking activities (yellow). Trade contractors return the signed original to you upon completion of the work. Using different colors makes the destination of each copy readily identifiable. Strive to issue work orders within one workday of the inspection.

Computerized Work Orders

Computer-issued work orders offer even more advantages. Consider the administrative time that can be eliminated by using wireless technology to forward hōmsoft™

September 05, 20__

HOMSoft Builders
39987 Main Street
Hometown, ST 55551
Phone (555) 555-5555
Fax (555) 555-5050

TO: <Trade Contractor>   Office phone   (952) 555-1212
       123 S. Main Street   Office fax   (952) 555-1234
       Hometown, ST 55551

Warranty Work Order

Dear Customer Service:

Please contact the home owner to resolve the following issue(s) as soon as possible. Report completed work to our office as soon as it is done.

Work Quickly, Professionally, and Courteously

Chuck Gundersen   **Home phone**   (555) 555-0000
4567 Harmony Lane   **Cell phone**   (555) 555-1234
Yourtown, ST 55553   **Closing date**   8/21/20__
Lot 6, Block 7   **Development**   Vista Park
Work order no. 19190   **Inspection date** 9/5/20__

Request no.	8	**Date assigned**	9/5/20__
Location	Living Room	**Days open**	0

Details   Inspect and water test roof above living room for possible leak. Water was reported coming through light fixture in center of room.

Correction: _____ By_____ Date_____

Home owner comment: _____ Initials_____

Thank you for your cooperation. Please notify our office promptly upon completion of these items, so we can maintain the most current information possible.

Sincerely,

HOMSoft Builders

work orders with the push of a button, while standing in the kitchen of the home you just inspected. In addition to improving response time, you reduce the possibility of errors that can occur when inspection notes are transcribed onto work orders in the office. Your system also should offer alternatives such as faxing or printing hard copies for any trades who still prefer older methods.

The trades may take computer-generated work orders more seriously than handwritten orders. To save time and errors, names, addresses, and phone numbers can be pulled up from the warranty database instead of repeatedly input to save time and errors. By coding each item to a trade and an item category, the raw material for detailed reports automatically accumulates. Such data might be sorted and analyzed by community, floor plan, trade, date, home owner, category of problem, and so on. Upon completion, enter the date and any notes into the computer to log out the work order.

Scheduling Warranty Repair Appointments

Builders typically work with three basic scheduling methods to accomplish warranty repair work. The one you use depends on circumstances and the personalities involved. Use the fourth alternative—that of accepting the home owner's key—with caution.

Work Date

The work-date approach has become popular in recent years. At the end of the warranty inspection, the warranty rep asks the home owner for a date to schedule the work. This date usually needs to be 10 workdays into the future to allow adequate notice to the trades. Put this work date on each work order. The home owner arranges access for that date, and all work can be completed simultaneously. Because work is consolidated on a known date, the warranty rep can stop in to check on progress and answer questions. Builders report a 75 to 80% success rate with this method. Repairs that require work to be performed in sequence (for instance, minor drywall work followed by paint touch-up) might be described on the respective work orders with letters, *AM* or *PM*, beside the work date to ensure the correct progression. More major sequencing issues require other scheduling methods.

Administrator Appointments

Another option is to have the warranty administrator communicate with the service technicians and the home owner. Companies that want to control scheduling for all work orders should recognize that this choice requires more staff members. If the administrator simply made one call to each party, the activity would still be time consuming. However, the warranty administrator seldom connects with both parties on the first try. Unfortunately, multiple contacts create many opportunities for messages to be tangled up or missed.

The administrator will need to know how long repairs will take to determine the next available appointment time for the in-house service technicians. This time can be difficult to estimate and may change once the technician begins the work. Repair appointments set too close together may result in the technician leaving before the work is complete and necessitating a second visit. Home owners hate this situation, and the technicians are not too crazy about it either. On the other

Two warranty technicians worked in a home while both home owners were present. The next day the husband called the warranty office asking the administrator to check with the technicians to find out if either of them had "accidentally" picked up a $1,200 gold wedding ring. Both technicians had been with the company more than 10 years and had impeccable records. Neither had seen the ring in question. Next day the home owner called again; the ring had been discovered on the floor under a dresser. Uncomfortable situations can arise even when adults are present during repair work. No system is flawless.

hand, repair appointments that are too far apart waste valuable time and wages, and they make home owners wait longer for needed repairs or other attention.

Direct Tech Contact

Having service technicians—whether in-house or trade employees—communicate directly with the home owners can also work. Tradespeople have a responsibility to cooperate in setting warranty appointments, and many do so effectively. Home owners also should feel welcome to expedite the process by initiating the call to the assigned trade— a good reason to forward paper or electronic copies of work orders to home owners. The technician who will do the work is in the best position to esti-mate the time needed. Direct conversation also provides the technician an opportunity to ask the home owner for additional details about the item.

Keys

A fourth alternative is to accept a key to the home. Remember that regardless of what the home owner signs in the way of a release, you are ethically (and in most states, legally) liable for problems that arise. This responsibility should never be taken lightly. Only individuals who are completely trustworthy should be sent into homes, whether or not the home owner is present. Although builders are increasingly reluctant to work in homes without an adult present, the alternative— setting appointments with people who work—can affect service response time dramatically.

If you offer a key release option to your home owners, obtain their written permission to enter their homes in their absence when you accept their keys, maintain total control of the keys and keep records of which keys you have and which ones you have returned. Your computer system should support this detail so staff can check the status of keys quickly and easily.

If you prefer to avoid accepting keys, your home owner guide should inform customers of this policy as in the example in Figure 6.2.

Repair Appointment Guidelines

To attain the level of service and consistent treatment you want for your home owners, set specific behavioral standards and establish policies covering repair appointments (Figure 6.3). Review this information with your staff and add any other appropriate points they suggest. Keep in mind that you can use the same information in working with your trades. Some of the points listed deserve a bit of

FIGURE

6.2 Home Owner Guide Entry: Warranty Procedures

Access to Your Home

[Builder] conducts inspections of interior warranty items only when an adult is available to accompany our representative and point out the items you have listed. Both our in-house service technicians and those of our trade contractors will likewise perform repairs only when an adult is available to admit them to your home. An adult is a person 18 years old or older who has your authorization to admit service personnel and sign completed work orders.

We do not accept house keys, nor will we permit our trade contractors to accept your key and work in your home without an adult present. While we recognize that this means processing warranty service items may take longer, we believe your peace of mind and security should be our first concern.

Pets

[Builder] respects the pets that many home owners count as members of their households. To prevent the possibility of an animal getting injured or lost, or giving in to its natural curiosity about tools and materials used for repairs, we ask that you restrict all animals to a comfortable location during any warranty visit, whether for inspection or warranty work. This policy is also for the protection of our employees and trades personnel. We have instructed [Builder] and trades personnel to reschedule the appointment if pets have access to the work area.

Personal Belongings

In all work that we perform for our home owners we want their personal belongings to be protected. When warranty work is needed in your home, we ask that you remove vulnerable items or items that might make performing the repair difficult. [Builder] and trade personnel will reschedule a repair appointment rather than risk damaging your belongings.

Signatures

Signing a work order acknowledges that a technician worked in your home on the date shown with regard to the items listed. It does not negate any of your rights under the warranty nor does it release us from any confirmed warranty obligation. If you prefer not to sign the work order, the technician will note that fact, sign the work order, and return it to us for our records.

additional explanation and that explanation appears in the following paragraphs. (The numbers correspond to the numbers in Figure 6.3.)

2. Work Dates. Review how the work-date method of scheduling is intended to function and where this information appears on the work order.

3. Confirm Supplies. Ensure that your service technicians understand the procedure for ordering materials. Options include having them tell the warranty administrator what they need or obtaining a purchase order number from accounting and

FIGURE

6.3 Home Owner Guide Entry: Warranty Repair Appointment Guidelines for Trade Contractors and [Builder] Tecchnicians

Preliminaries

1. Check the work order to see if you have any questions; if so, contact the warranty administrator for help.
2. If the work order includes a work date, make note of it on your calendar.
3. Confirm that you have all necessary tools and materials, or order required items.
4. As often as possible, estimate the amount of time the work will take and arrange your schedule to complete the repair in one trip
5. If no work date appears on the work order, contact the home owner to schedule a repair appointment.
   - Give yourself a 30-minute range to allow for traffic and other unexpected events (for example, "between 8 and 8:30 a.m.")
   - Minimally commit to an a.m. or p.m. time frame.
6. If you will be late, call as soon as you know, apologize, and offer to reschedule.

Upon Arrival

7. Park in the street.
8. Refrain from smoking on the home owner's property.
9. Take the work order and a business card with you to the door.
10. If you have not met the home owner before, introduce yourself and give him or her a business card.
11. Avoid using first names until the home owner invites you to do so.
12. Remove or cover your shoes before entering the home.
13. Review the repair item(s) and their location(s) with the home owner if necessary.
14. If the home owner changes or misses the appointment, note this on the service order and inform the warranty administrator.
15. If the home owner is not home when you arrive:
    - Call the phone numbers provided on the work order. If you are unable to reach the home owner, wait a minimum of 10 minutes and then leave a door hanger so the home owner knows you were there.
    - Notify the warranty administrator.
    - Contact the home owner to reschedule.
    - If this appointment is the second one the home owner has missed, the warranty administrator should schedule the next appointment and follow up to confirm it.
16. While an appointment may not be necessary, contact the home owner prior to performing any exterior work. After completing exterior work, leave a door hanger so the home owner knows you were there.

FIGURE

6.3 *Continued*

Repair Work

17. Check the work area. If any cosmetic damage exists, note that damage on the work order and ask the home owner to initial your note.
18. Be aware of your surroundings to ensure that children, pets, and the home owner's belongings will not be harmed.
19. Protect counters, floors, and other surfaces from tools, tool boxes, drips, and dust with a drop cloth or other material.
20. Ask the home owners to remove personal items or furniture from the work area.
21. If conditions are inappropriate for the work you need to do, leave and reschedule the repair appointment.
22. If you need to use the home owner's electricity or water to perform the repair, ask permission first. Avoid using the home owner's phone or bathroom unless the home owner volunteers that you are welcome to do so.
23. Perform the needed repair(s) efficiently and within [Builder] standards.
24. If you are uncertain about how something should be done, look at the model home or contact the warranty manager.
25. If you need to make a follow-up visit to complete the work, schedule it immediately and inform the warranty administrator.
26. Refrain from making any negative comments to home owners about the work of others, the company, its personnel, or the community.
27. If you notice a serious problem, report it to the warranty office immediately.
28. Use the 10-minute rule to screen new items. Correct the item if you know it is covered by the warranty, you have the necessary material, and the work can be completed within 10 minutes. Note the item on the work order so the warranty file is complete.
29. Immediately report any damage caused by tool boxes, dirty shoes, mistakes, or accidents so the company can arrange for appropriate corrections.

Conclusion

30. Clean up the work area: remove all tools, materials, dust, and debris.
31. Make brief notes about what you did to repair the item(s).
32. Inform the home owner when you are finished.
33. Request the home owner's signature on the work order. If the home owner is unavailable or prefers not to sign the work order, make a note of that fact on the work order.
34. Sign and date the work order.
35. Turn the work order in to the warranty office.

ordering items themselves. Some companies set up charge accounts at local suppliers or issue a company charge card.

4. One Trip. Another point that needs continuous reinforcement is the value of completing repairs in one visit. By emphasizing the goal of one-visit repair when hiring trades and warranty personnel, the builder establishes his or her priorities.

Effective repair appointments require accurate diagnosis and sufficient details; therefore, the technical knowledge of builder field reps is critical. With the right information, the trade knows exactly what to do, what materials and tools to bring, and how much time to schedule in the home. The warranty work order must then include all of these details. Back up these efforts by having staff readily accessible to repair personnel to answer questions that may arise during the repair work.

As a safety net, support your one-visit repair policy with procedures for managing multiple-visit repairs (see Chapter 7). Some repairs legitimately involve several steps by one trade or require the efforts of several trades in sequence. Explain the steps to the home owner and accept responsibility for the close administrative attention to move trades through the home in rapid succession.

5. Appointments. Expecting a home owner to stay home all day waiting for a 10-minute repair that finally gets done at 4:30 p.m. is unreasonable. Expect appointment times to be set up within—at most—a four-hour block. Anyone should be able to keep the first appointment of the day and possibly the first appointment immediately after lunch time. Put a lot of emphasis on calling if you are delayed. In spite of the proliferation of cell phones, this common courtesy seems to be extremely uncommon with some trade contractors.

18. Pets. Make the company's position on pets clear by covering the subject in your home owner manual. To accomplish this task, you may want to use the wording for this topic in Figure 6.2.

20. Home Owner's Possessions. Likewise, clarify the company's position on who removes personal belongings by discussing that subject in your home owner guide as well. Again, Figure 6.2 includes one approach to this policy.

21. Inappropriate Conditions. Discuss what constitutes "inappropriate conditions." Service techs encounter everything from snarling dogs to lonely women in negligees. Suppose an interfering relative is visiting? If a home owner engages in harassment and criticism, the tech may decide simply to leave the home and report the problem to the warranty manager rather than permit an argument to develop. While you are unlikely to anticipate every scenario, these discussions are an excellent exercise in developing the judgment skills of your staff.

Once you agree to perform a repair, even one in the infamous "gray area," do it in a first-rate manner. No one who takes customer satisfaction seriously should ever say, "I did them a favor by agreeing to the repair in the first place, and now they want it done right as well."

23. Repair Quality. The so-called Band-Aid approach to repairs generally fails. Saving money with shortcuts or halfway measures is a false economy. Recognize that the item in question caught the home owner's eye, and the repair or correction you provide is likely to be carefully evaluated.

26. Negative Comments. Relate examples of incidents where negative remarks escalated problems. For instance, a landscaper, planting a shrub, was questioned by the home owner about using soil amendments in the hole. The landscaper replied, "Yeah, that'd be the right way to do this, but the builder won't pay us to do it right. This is what you get."

28. Additional Items. Well-trained technicians can screen new items pointed out by the home owner with the 10-minute rule: Correct the item if you know the item is covered by the warranty, you have the necessary material, and the work can be completed within 10 minutes. Note the item on the work order so the warranty file is complete. For items that would take longer than 10 minutes, require parts not on hand, or on which warranty coverage is questionable, refer the home owner to the normal warranty reporting process.

29. Damage. A service tech may be hesitant to volunteer that he or she damaged a surface or a home owner's belonging. Assure your staff that infrequent accidents of this type will not jeopardize their jobs, but failure to own up to such incidents might.

31. Notes About Repair. "Done" is not usually what you want for notes, but unless you discuss the notes you do want with the service staff and your trades, "Done" is what you are likely to get. Show repair staff and tradespeople examples of work orders with appropriate notes. If technicians turn in work orders with insufficient information, return them to the technicians and request more detail.

33. Home Owner Signatures. Work order forms include a line for the home owner's signature to indicate satisfaction with the work or at least to acknowledge that work was performed. Whether or not this requirement accomplishes anything positive in a service program is open to discussion. This signature guarantees neither home owner satisfaction nor release of the company from further obligation. If a husband signs a work order, but his wife is displeased, the builder and the trade still have a problem. If both home owners work, and you've gained access by using their key, the practical obstacles to obtaining a signature can cause a genuine inconvenience to both the home owner and the trade.

Warranty technicians sometimes hold work orders waiting to catch folks at home until long after the work is complete. This situation creates an inaccurate picture of outstanding work. Further, by requesting a signature, the technician gives the home owner the authority to set standards. Difficult home owners may try to take advantage of this opportunity to leverage their positions. An entry for your home owner guide similar to the one in Figure 6.3 can forestall some of this delay.

Instruct service personnel to obtain the home owner's signature, if possible, but if that signature is unavailable (the home owner is not there or refuses to sign for some reason), they should make note of the circumstances on the work order, sign it, and turn it in immediately rather than hold it.

Training Activities

You are unlikely to anticipate every possible repair scenario. However, you can make great strides using experience and common sense. By discussing previous situations, staff will be prompted to recall and suggest others.

Warranty Repair Appointment Guidelines

Working with your in-house technicians and available tradespeople, review the points in Figure 6.3. Discuss examples and situations; then share experiences and insights. In particular, role play the conversations suggested by items 10 (introduction), 16 (exterior work), 20 (personal items), 21 (inappropriate conditions), 22 (permission to use utilities), and 33 (signature). Over time, you will no doubt modify the list of warranty repair behaviors based on sound suggestions from your team and their experiences.

Workshops with Trades and Suppliers

Monthly or quarterly workshops conducted by trades or suppliers combined with visits to their showrooms or shops can be helpful for repair technicians. Most trade contractors are happy to provide minor parts and materials to in-house warranty technicians in exchange for avoiding a trip to a home to do a minor repair. As a side benefit, supporting the trades in this manner often leads to them being more responsive when significant work is needed.

Special Handling

While the value of clear procedures for handling normal warranty items cannot be overstated, sometimes special situations require even more detailed methods. This chapter offers suggestions for (a) responding to some of the most common situations and, where appropriate, (b) searching for their causes. Because your warranty work may include a subject not addressed here, the first special handling procedure is a generic analysis that can help you untangle an unusual situation.

New Warranty Challenges

Along with routine warranty matters, nearly every builder has some warranty baggage and at least one unusual warranty storm waiting on the horizon. These situations vary from that unexplained foul odor that emanates from Mrs. Jones's disposal, to the mysterious thump that occurs between midnight and 3 a.m. in the Hernandez's attic, to the leaky windows your company installed in 37 homes.

Identify your warranty baggage by finishing the sentence, "This job would actually be easy if we could just resolve . . . ." Rest assured that when you resolve that issue, a new one is likely to take its place. Storm signals include getting a second call about the same complaint—usually defective material—within a short period of time. Another warning sign is hearing about other builders who are experiencing difficulties with a material or method—and knowing your company has used it. New problems need new solutions.

Working in unfamiliar territory is uncomfortable. If you are unsure of what to do, that state of mind usually means you need more information. A systematic approach to unusual situations reestablishes control and confidence. Figure 7.1 provides a checklist to support your efforts. Designate one person to oversee this effort; that person may well be you. Emphasize documentation at every step and, when appropriate, set up a reporting schedule for regular updates to management.

Intranet-Based Procedures

Having procedures and company forms available to all employees via a company intranet is both convenient and logical. That structure may be most useful in special warranty situations. Warranty reps in the builder's communities should each

FIGURE
7.1 Special Handling Analysis

1. Define and Measure the Problem

Number of homes affected?
Specific community or model(s) affected?
Physical effects on the home?
Emotional impact on the home owners?
Financial impact on value of property?
Correction costs?
Prevention costs?

2. Identify Resources and Partners

List the individuals, agencies, or companies who might assist you in achieving a resolution, either through information, funding, or doing the actual work.

3. Examine Product Quality

Take a brutally honest look at what your company's role in the situation is. What, if anything, are you doing (or not doing) that might be a factor? Consider the following factors:

> Design
> Materials
> Installation
> Sequence of construction
> Supervision

4. Review Expectations You Set with Home Buyers

What do your buyers hear or read as they progress through your process that applies to this situation? Where in your process would you appropriately inform your home buyers about this subject?

A. Conversations

- Conversations customers have with sales and other personnel
- Preconstruction conference agenda
- Frame stage tour agenda
- Orientation agenda
- Warranty appointments

B. Documents

- Contract
- Disclosures and riders
- Warranty and warranty standards
- Maintenance guidelines
- Manufacturers' literature

FIGURE

7.1 *Continued*

5. Establish a Response Procedure

Determine and prioritize actions you need to take for the following three groups of customers:

A. Home Owners Who Have Complained

- Inspection checklist to assure consistency, no matter who performs the inspection
- Repair options and recommendations
- Detailed documentation such as work orders and follow-up letters or reports

B. Future Home Owners

- Conversations with sales and other staff
- Documents of the sale
- Home owner guide
- Standard meeting agendas
- Warranty appointments
- Trade personnel conversations

C. Pending Home Owners

These folks have not closed yet, but they have signed their contracts. You may need a buyer's update before or at closing so you can stop problems before they start. This step is an interim task, necessary only with those buyers who are already in your system.

6. Follow Through with Training

Avoid assuming that because department heads got regular updates, this information has filtered through to frontline personnel. Ask yourself what your personnel need to know and figure out how they will learn it.

adhere to standardized protocols for unusual situations. These events seldom can be well-handled with a yellow tablet. While you can keep everyone on the same page with hard copies of checklists, forms, and procedures, the practical difficulties make success elusive, and you continually run the risk of having a warranty rep respond to a situation or home owner with old material.

Water Intrusion

Water intrusion expert Russell Nassof of TRC/Environomics advises builders to have five elements in place for successfully managing these undesirable events. With these five elements you will enhance your company's ability to reduce risk and liability associated with water intrusion and mold issues. Your best chance of addressing such occurrences and retaining customer goodwill is careful planning.

Prevention

Carefully design and construct your homes and use quality control practices. Use strong preventive measures at all stages of construction—starting at the earliest planning, preconstruction, and design phases of a project and continuing through project completion. You may want to include third-party inspections as an extra precaution.

Response

Your reaction must be prompt, empathetic, and comprehensive. Give water intrusion emergency status and take immediate action. Conduct a thorough inspection—including generous use of a digital camera—to identify the source of the problem and the significance to the home. Builder checklists guiding this process are typically many pages to ensure no detail is overlooked and all conditions are documented.

Be candid with the home owners and relocate them during remediation. Follow established best work practices during repairs including containing and dehumidifying the affected area. If someone else is responsible for the situation, communicate that quickly with detailed documentation.

Organization and Training

Designate a staff member to be trained specifically to respond to water intrusion events. Train frontline personnel—including salespeople—about what to say, what not to say, and appropriate remediation and response procedures.

Risk Transfer

Ensure that your trade agreements, purchase agreement, warranty, and home owners guide all contain clear descriptions of the responsibilities of the involved parties. Accountability is easier to establish if parameters are in writing and agreed to prior to an emergency.

Documentation

Documentation, including photographs, test and inspection data, field notes, and interviews, is critical in these situations for several reasons. First, having a clear procedure to follow impresses the customer with your competence in handling the issue. To reach an appropriate diagnosis and response, gather the facts systematically. Reflecting on the data may lead to improvements in quality that will prevent future problems. Moreover, if you need to defend your decisions or actions, thorough documentation is your best ally.

Roof Leak

Consider the example of a roof leak—an event practically guaranteed to make home owners frantic. Roof leaks conflict with one of the fundamental purposes of a home—to protect the occupants from the weather. Besides following planned response protocols, provide a candid explanation to the home owner, including the following points:

- For safety reasons and to ensure the effectiveness of the repair, typically repairs can be made only when the roof is dry.
- The repair order will be given top priority for the next dry day.
- You can feel certain that the repair was successful after Mother Nature tests the repair several times.
- <u><name, phone number, and e-mail for contact person></u> will contact you after the next several rains to confirm the success of the repair.

This level of attention is easier with a simple reminder system. The format shown in Figure 7.2 works well. After three successful follow-up calls, remove the home owner from the call list. Again, you could track this repair on your computer system.

Wet Basement

As with roof leaks, another problem that can send home owners into a frenzy is a wet basement. These problems also share the annoying condition that, until rainy weather has tested the repair, no one knows whether the problem has been solved. Confidence in a repair increases as time passes and the work is tested.

When a wet basement report comes in, respond promptly, and inspect the home as soon as possible. An inspection sheet such as the one in Figure 7.3 documents conditions.

If home owner actions are the cause, suggest appropriate corrections and then offer to inspect those corrections when the home owner has made them. Document your recommendations in a follow-up letter.

FIGURE 7.2 Roof Leak Follow-Up

[Logo] Roof Leak Follow-Up

Leak date	Home owner	Phone and e-mail	Repair date	Dates of follow-up calls		
7/18	Ornstien	555-440-8924 bornstein@att.net	7/20/20–	7/24	8/1	8/17
8/1	Raymond	tanger@aol.com	8/3	8/17	8/24	9/2
8/25	Everett	johne@earthlink.com 555-461-9440	8/28	9/2	9/21	9/24
10/6	Smith	555-440-6529	10/10	10/20		

FIGURE

7.3 Wet Basement Inspection

[Logo]

Wet Basement Inspection

Purchaser _Cindy Buyer_   Date _7/16/20–_   Time _2:00 pm_

Address _3399 West Port Ave_   Community _Bailey's cove_

Home phone _555-555-1234_   Home site no. _3-27_

Work phone _n/a_   Plan _Trinity Bay_

Cell phone _555-702-0050_   E-mail _____

House File

- ☑ Grade plan
- ☑ Final survey
- ☑ Quality management inspections
- ☑ Orientation list
- ☑ Warranty document and guidelines
- ☑ Previous warranty requests
- ☑ Previous inspection reports
- ☑ Phone logs
- ☑ Correspondence

Existing Grade and Landscaping

- ☑ Photos
- ☑ Slope at foundation
- ☑ Landscape materials in backfill area
- ☑ Edging
- ☐ Downspout extension and splash blocks
- ☑ Location of trees or large shrubs

- ☐ Sprinkler system
- ☑ Water service
- ☑ Drainage from neighboring lots
- ☑ Home owner additions? _None_

Interior Conditions

- ☑ Foundation cracks
- ☐ Floor cracks
- ☐ Crawl space condition
- ☑ Penetrations (form ties, utility lines)
- ☑ First entry location
- ☑ Area affected
- ☑ Perimeter drain
- ☑ Sump pit
- ☑ Sump pump
- ☑ Water lines
- ☑ Sewer lines
- ☑ Condensation
- ☐ Home owner basement finish?

Notes

As at 60-day, downspout extensions are in the up position; roof water is flooding backfill soils.

Explained to Mrs. B again about imporance of keeping these down to channel roof water away from home. Sprinkler system installed by home owner—appears ok. Minor floor cracks and crawl space are dry. Mrs. B intends to finish B/M.

Home Owner _Cindy Buyer_

Home Owner _____

Builder _Harry Beasley_

Defective Material

A sinking feeling accompanies the realization that your company installed defective materials in a number of homes. As you investigate, keep in mind that material defects can result from the following causes:

> Having a clear procedure reassures the home owners that someone who knows what to do is taking charge. This reassurance increases the home owner's confidence in actions you take or advice you give.

- The component has design flaws; the entire product line has the defect.
- One batch of the product turns out to be defective.
- Errors occurred during installation.
- The home owner improperly maintained the material.

To minimize such catastrophes, take the following steps:

- Select components with long-term performance in mind.
- Conduct research to learn about products.
- Avoid basing purchasing decisions solely on cost; saving a little now can cost a fortune later.
- Review manufacturer's installation instructions in detail and follow them.
- Consider having a manufacturer's representative inspect and sign off on installations until crews are proficient. (Many manufacturers offer training seminars or provide training videos.)
- Maintain accurate and complete files so you can determine what materials you used and where.
- Include appropriate maintenance information in your home owner guide.

If you find yourself dealing with defective materials in two or more homes, use the analysis in Figure 7.1 to come up with a comprehensive plan for responding. Figures 7.4 and 7.5 show examples of correspondence and questionnaires for home owners who might be affected.

Structural Warranty Claims

As confusing as new home warranty coverages are, probably the most serious confusion surrounds structural coverage. Because a home is a structure, structural warranty coverage is sometimes assumed to mean things it does not mean—by salespeople and customers.

Prevent Confusion

Confirm that your salespeople understand the difference between the materials and workmanship warranty and the structural warranty. Rehearse conversations they might have with buyers on this topic. Clearly written warranty documents must support the efforts of company salespeople. Make these documents available for home buyers to review prior to purchase and deliver copies to them with their contracts.

FIGURE
7.4   Material Defect

Dear <Home Owner>

Several [Builder] home owners have contacted us regarding their concerns about windows, including water leaks and what may be excessive air infiltration.

As a result, we are conducting an investigation to determine whether the problems experienced are caused by an error in installation or a faulty product.

If the installation is the cause, [Builder] will correct this. If the product is faulty, we will work with the supplier and manufacturer to correct the concerns reported. We have initiated a dialog with the supplier and the manufacturers about this matter.

In a situation of this type, good data is essential. We need a complete list of which homes and which windows may be involved. You can assist in this effort by providing us with information about your home.

Please use the Window Survey form attached or enclosed to alert us to any concerns you have about your windows. Some air and even dust normally infiltrate windows, especially during strong winds or while the area around your home remains unlandscaped. However, excessive air or any water is unacceptable.

If you indicate any concerns on the attached or enclosed form, we will contact you for an inspection and will keep you informed regarding our findings and actions. You can expect an update on the response from the supplier and manufacturer within the next several weeks.

Your assistance with this investigation is appreciated.

Sincerely,

Warranty Service Manager
[Builder]

Attached or Enclosed: Window Survey Form

Response to Structural Claims

Signs of structural movement or failure of some load-bearing portion of a home usually appear in a pattern. Damage might involve cracks in the foundation, uneven floors, cracks in drywall, or ill-fitting doors and windows. Individually, each of these can have an innocent explanation; in combination, they require aggressive attention. On the other hand, home owners sometimes interpret the typical effects of settling as structural damage and panic. Either way, prompt response can mitigate the negative effects. The underlying cause can come from one or more of the factors listed below:

FIGURE
7.5 **Window Survey**

[Logo]

Window Survey

Purchaser _Cindy Buyer_   Date _8/10/20–_

Address _____   Community _Bailey's Cove_

Home phone _____   Home site no. _____

Work phone _____   Plan _____

Cell phone _____   E-mail _____

☑ The windows in my home are performing satisfactorily.
☐ I have the following concerns about windows in the following rooms and would like an inspection:

_____ _Thanks for asking—we are_ _____

_____ _fine, no window concerns._ _____

_____

_____

_____

Please note that condensation is caused by humidity
within the home and is not considered a leak.

Home Owner _Cindy Buyer_ _____

Date _8/10/20–_ _____

* The engineer who designed the structural system, the builder who built it, and the home owner who owns it can each do everything right. However, once again, Mother Nature gets the last word.
* In spite of checks and balances during construction, builder errors can occur. Plans and specifications may not have been followed correctly or one or more of the materials used may be defective.
* The home owner may alter the home or drainage surrounding it and, in turn, cause the problem.

Preliminary inspection of reported symptoms is appropriate prior to getting an engineer involved. Paying an engineer to diagnose nail pops and shrinkage cracks is wasteful. Instead, first use a checklist similar to the one shown in Figure 7.6 to document conditions at the time of your inspection and determine appropriate action. If you suspect structural damage, schedule an engineer's inspection.

FIGURE

7.6   Structural Inspection

[Logo]

Structural Inspection

Home Owner _Kyle Everett_   Date _6/14/20–_ Time _1:30 pm_

Address _3741 Seashell Court_   Community _Bailey's Cove_

Home phone _555-702-1886_   Home site no. _4-11_ Plan _Sunrise_

Work phone _555-702-9426_   Plan _6/10/19–_

Cell phone _555-440-0026_   E-mail _hylee@earthlink.com_

Review House File

- ☑ Previous report of concerns _none_
- ☑ Grade certification
- ☑ Perimeter drain included in foundation plan _yes_

Existing Grade and Landscaping

- ☑ Changes to original grade _none_
- ☑ Slope at foundation _good_
- ☑ Landscape materials in backfill area _ok_
- ☑ Edging _ok_
- ☑ Downspout extension or splash blocks _ok_
- ☑ Location of trees or large shrubs _NA_
- ☑ Sprinkler system _NA_
- ☑ Water service and meter _NA_
- ☑ Drainage from neighboring lots _NA_
- ☑ Home owner additions to exterior _none_

Foundation

- ☑ Wall cracks (size, location, mark with 2″ line & date) _none_
- ☑ Bowing _none_
- ☑ Sill plate/grout _none_
- ☑ Beam pockets _none_
- ☑ Floor cracks (location, width, displacement) _none_

- ☑ Heaving (note furnace collar, flexlines, slip joints on drain lines, basement stairs) _none_
- ☑ Condition of floor in relation to support posts _none_
- ☑ Moisture or standing water, including crawl space _none_
- ☑ Home owner construction in basement _none_

Interior

- ☐ Drywall
- ☐ Doors
- ☑ Floors out of level
- ☑ Trim boards, cabinets, and so on
- ☑ Masonry or tile work
- ☑ Windows

Exterior

- ☑ Concrete flatwork
- ☑ Patio covers
- ☑ Masonry
- ☑ Trim boards

Photos taken?   ☑ Yes ☐ No
Reshoot grades?   ☐ Yes ☑ No
Engineer required?   ☐ Yes ☑ No

Continued

FIGURE
7.6 Continued

Notes

_____

_____ *Steel beam raised slightly—need to adjust post to level; repair dry-* _____

_____ *wall crack and adjust doors after beam is level. Kyle has applied cus-* _____

_____ *tom color paints and will do his own touch up after work is complete.* _____

_____

_____

_____

_____

Home Owner _____ *Kyle Everett* _____

Home Owner _____

Builder _____ *Harry Beasley* _____

Your engineer may need to check the home more than once to discern a pattern of movement. Once the cause is determined, the engineer can outline corrective measures. After corrections are implemented, a waiting period typically follows. Confirming that the home is stable makes sense before performing cosmetic repairs; this practice can prevent the frustration of doing them twice.

This process—inspect, wait, inspect, repair, wait, repair—can span several months and easily exceed the limits of your home owner's patience. For best results follow these guidelines:

* Empathize with the double stress that the home owner experiences: today's inconvenience and perceived threats to tomorrow's value.
* Maintain regular communication. Stay in touch at every step.
* Outline what will happen and why.
* Answer questions with candor and define construction jargon as needed.
* Document all meetings, phone calls, schedules, and agreements carefully and promptly.

Severe Weather

When storms (or any natural disasters) strike, home owners may assume their builder should repair any damage. This case provides an example of being first to raise a subject, which bestows advantages. A mass communication directing home owners to their insurance carriers for help can dramatically reduce their phone calls to (and arguments with) your warranty office staff.

Imagine a severe wind storm hits your area. The next day, send e-mail notices similar to the one in Figure 7.7 to your home owners. Print out labels for all home owners from your database, and mail them a similar letter. Post the update on your Web site as well. When the phone calls start, your staff has a positive response. "Yes, Mr. Patel, that was a terrible storm. In fact we are sending our home owners guidelines on how to proceed. May I fax you a copy right now?" This response is better than, "No, we don't cover storm damage. Call your insurance company." Pointing out to your caller that the company is sending this same answer to all home owners normalizes the information and makes the message easier for the caller to accept.

Multiple-Trade Repairs

Repairs that involve several trades working in sequence require more attention than single-trade repairs. The work-date approach to scheduling described in Chapter 6 is probably inappropriate for multiple-trade repairs. By establishing a routine method that supports the extra attention these repairs require, you stay on top of the work.

The classic example of a multiple-trade repair is the plumbing leak that has damaged drywall and, therefore, requires up to three trades to complete the

FIGURE
7.7   Severe Weather

Dear <Home Owner>:

Yesterday's storm was severe and according to news reports caused significant damage to trees and homes in our area.

We suggest that you promptly inspect your property and report any damage you find to your home owner's insurance company. Look for pieces of shingles in your yard or gutters, broken tree limbs, damaged fencing, or other effects of the storm.

Photographs help document such damage and may support your claim. Although storm damage is excluded from our limited warranty coverage, if you need information on how to prevent additional damage while you await an inspection and response from your insurance company, we may be able to make preventative suggestions.

If you have questions, please contact me.

Sincerely,

Warranty Service Manager
[Builder]

repair. The first visit usually occurs promptly. But often nothing further happens until the home owner becomes angry enough to call the warranty office. To avoid this excitement, identify all the steps needed in sequence. Estimate how long each portion of the work should take and issue the work orders with that information. Remember to consider whether you should include a cleaning crew in your plan. Look to your computer to assist you with the close and necessary monitoring of such repair sequences. Figure 7.8 shows an example of computer tracking for multiple-trade repairs.

You can also track this series of work orders by hand on a multiple-trade repair work sheet similar to the one in Figure 7.9. This tickler can remind you to check progress daily and confirm the next appointment until the work is completed. It also provides a place to document any changes in the schedule that occur along the way—whether they are caused by a trade contractor or a home owner.

When working with home owners on this type of repair, outline the repair steps. Check in with the home owner as you mark each step completed. Steady progress combined with the assurance that someone is monitoring the work usually will keep the customer satisfied. Alert your trades to the circumstances to gain their cooperation.

FIGURE

7.8   Tracking Multiple-Trade Repairs Electronically

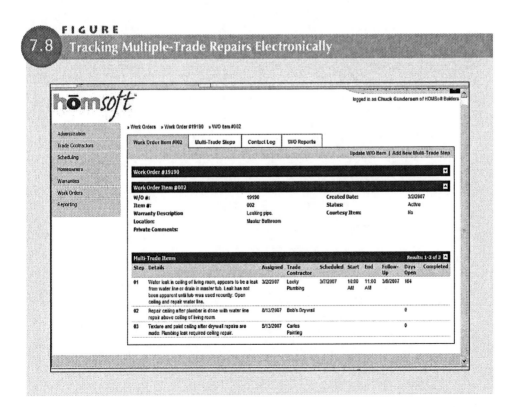

FIGURE
7.9   Tracking Multiple-Trade Repairs Manually

Home owner _Aaron Ornstein_   Date _7/20/20–_

Address _617 West Port Lane_   Community _Bailey's Cove_

Home phone _555-440-8924_   Home site no. _3-41_

Work phone _works from home_   Plan _Port View_

Cell phone _555-702-6112_   E-mail _bornstien@att.net_

Confirm appointment	Work order to	Confirm completion	Back charge?
✓	Trade _Joe's Plumbing_ Phone _555-555-9999_ Work order no. _7519_ E-mail _joe@plumbing.com_ Scheduled time _8:00_ (a.m.)/p.m. Date _7/21/20–_ Notes _Aaron will call when work is complete._	Done 7/21/20– per Aaron	Yes
✓	Trade _Dale Drywall_ Phone _555-702-1122_ Work order no. _7520_ E-mail _dalez@wall.com_ Scheduled time _10:00_ (a.m.)/p.m. Date _7/22/20–_ Notes _They will need 2 visits to complete; Aaron intends to call when work is finished._	Done 7/24/– per Aaron	No
✓	Trade _USA Painting_ Phone _555-440-1617_ Work order no. _7521_ E-mail _john@usapinting.com_ Scheduled time _8:00_ (a.m.)/p.m. Date _7/28/20–_ Notes _John will confirm completion._	Done per John 7/28 7/28	No
✓	Trade _Classy Cleaners_ Phone _555-702-6543_ Work order no. _7522_ E-mail _maggie@aol.com_ Scheduled time _4:00_ a.m./(p.m.) Date _7/28/20–_ Notes _check w/ Aaron when complete_	Done – per Maggie. Aaron is very pleased.	No

Turnover of Common Area to Home Owners Association

The turnover of common areas to a home owners association marks another significant turning point in a builder's relationship with a community. As with the expiration of the material and workmanship warranty for individual home owners, this turning point creates a separation to which the builder looks forward and which the home owners association usually resists. A deliberate strategy for managing this rite of passage helps. Prepare a clear overview of your procedures and deliver it to your buyers with other home owner association documents. Behind the scenes, include a discretionary amount in your budget from the beginning so that you will be prepared to provide something extra.

Develop a checklist of common area items to inspect—sod, shrubs, trees, sprinkler systems, post lamps, concrete flatwork, rock walls, and so on. Support the list with appropriate standards. Set a timely appointment to conduct an association orientation on the common areas.

To avoid multiple conversations and resulting misunderstandings, invite no more than three association representatives. Include at least two company people—one to talk and one to record decisions. Invite appropriate trade representatives—sprinkler, pool, and so on. With the trade contractors' help, review proper use and maintenance of each component, turn over any manufacturers' literature, and identify items to correct, repair, or install. Provide the association representatives with a copy of the list you compile.

Make this list a top priority and promptly complete each item you listed. Expect that during this process, the home owners association may ask for something extra. Use your discretionary budget amount to be a hero and pleasantly make a point of the gift. Upon completion of all items, send a letter similar to the one in Figure 7.10 to the association, with a copy to all affected home owners.

Community Close Out

When builders are involved for several years in construction of large communities, setting up satellite warranty offices within the communities themselves often may make sense. Refer to the discussion of satellite offices in Chapter 14 for additional guidance. Eventually, this office will be closed and the personnel assigned to new locations. Anticipate that this chain of events will capture the attention of the home owners.

Preempt difficulties or disappointment with proactive communication. Several weeks in advance of removing staff (furniture, the trailer, or whatever) send a letter to home owners alerting them to this change and providing them with clear directions on how to obtain answers to questions or report warranty items in the future. For instance, if the phone number or e-mail address they are accustomed to using is changing, provide a refrigerator magnet with the new contact information.

FIGURE

7.10 Turnover of Common Area

Dear <Board of Directors>:

Congratulations! The common areas in filings 3 and 4 at <Community> are now under the control and care of the <Community> Home Owners Association.

The list we created in our tour of the common areas on <date> is now complete. Maintenance of these areas and all amenities they contain is now the responsibility of the Association.

In addition, as you requested, [Builder] purchased and installed two trees in the entry island and added more ground lighting for the entry signs. The results are beautiful, especially at night. We are pleased to present these items to the Association as a gift.

All information, warranties, instruction manuals, supplier and trade names, addresses, and phone numbers have previously been turned over to the Association. < Pool contractor> will be in touch with you to finalize the pool maintenance agreement. We have been satisfied with their service, and I think the Board has made a wise choice to retain them.

If other questions regarding the care or operation of any mechanical features arise in the future, please contact me.

Sincerely,

CEO or President
[Building Company]

Nonwarranty Concerns

Home owners (and sometimes staff members of other departments) often contact the warranty office directly with *nonwarranty* concerns. The nonwarranty situations listed below provide examples of the strange issues that sometimes come to the warranty office:

* The tennis court that the developer was supposed to have installed last summer still doesn't exist.
* The county neglected to replace the stop sign after a car wreck took it out.
* A computer glitch overcharged four home owners for property taxes at their closings.

Warranty personnel often view this reality as unfair and burdensome. After all, they reason, "We're here to do warranty; why are they throwing this at us? It's not our job. Why are we the dumping ground?" This frustration comes from having plenty of other work to do and sometimes from the warranty staff not knowing themselves where to look for answers.

For the sake of the reputation of the warranty office, builders need a healthier approach. Begin with a can-do attitude. Look upon strange situations as opportunities to learn something new and to help. See the warranty department as the clearing house for issues that need direction. When a dumping-ground issue comes to the warranty department, follow these steps:

> When warranty investigates and conveys information, home owners will not need to use trial and error and repeat a story over and over until they connect with the person who can help. This aspect of seamless service provides a desirable situation for home owners (or staff members) who need help. With follow-up, this seamless service makes a powerful impression.

* Take ownership of the issue.
* Find the responsible party. (Use the Special Handling Analysis form in Figure 7.1 as a guide.) What you learn is often useful in future situations.
* Report the situation to the responsible party.
* Update the individual who contacted you about the issue. Explain what you learned, what steps you've taken, and who will handle the issue from here forward.
* Follow up a few days later to confirm that progress is being made. This attention means a great deal to the original complainer.

Consumer Protection Entity

In a behavior called *system skipping,* home owners report warranty items to the Better Business Bureau, a licensing board, or other consumer protection entity without ever having reported them to the builder. If this situation occurs, Inform the consumer protection entity about the facts and your response with a letter similar to the one in Figure 7.11. Avoid long-winded descriptions about how impossible these home owners have been since day one. (Save it for your therapist.) You may also want to call the home owners and review normal reporting procedures. Assure them that by following those procedures, you can provide the needed warranty work faster.

Disagreement with Neighbor

When they imagine their new homes, few home buyers envision disagreements with their future neighbors. However, such situations can occur, and often a home owner's first action is to take the problem to the builder. When this situation happens, the letter in Figure 7.12 can help you put the situation back in perspective for the concerned home owner.

When these arguments between neighbors involve drainage, the builder's best interests involve getting the matter effectively resolved because water draining from a neighbor's yard can have an impact on the foundation on an adjoining

FIGURE
7.11   Consumer Protection Entity

Dear <Name of contact person>:

Your notice number <identifying case number> regarding <home owner> arrived on <date>, and we contacted <home owner> on the same day to schedule an inspection appointment.

The inspection occurred <date>. As a result, [Builder] has issued three work orders to appropriate trade contractors and one to an in-house service technician (copies enclosed). All work listed should be complete by <date>.

We found one item listed by the home owners to be a normal home maintenance item, service for which is excluded from our warranty. We have informed the home owners in writing of this fact.

A review of our file for this home established that the home owners had not previously reported any of these items to us for attention. Therefore, we have reviewed normal reporting procedure with the home owners in hopes of expediting any further warranty repairs needed for their home.

We appreciate your attention to this matter. If you have any questions, please contact me.

Sincerely,

Warranty Service Manager
[Builder]

---

lot. Avoiding the tangled mess and finger-pointing that can result is best. Solid information about landscaping responsibilities backed by reasonable home owner association rules can minimize such situations.

The best protection a builder can have is documentation of the grades that existed at the time of delivery. Companies are sometimes reluctant to invest in such documentation. Later, they learn that this choice was a false economy when they spend significant amounts correcting problems they did not create. In a disagreement with a home owner, what you know is little help unless you can prove it.

Disgruntled Home Owners

Some home owners want to damage your company's business so badly that they are willing to sacrifice the value of their investment to do so. With signs or banners, they announce to the world, "This is a bad house. Talk to me before buying from [Builder]."

Begin with a direct approach. Confront the home owner. "Mrs. Jones, I see by the search light and digital readout sign in your front yard that you are unhappy.

FIGURE

7.12 Disagreement with Neighbor

Dear <Home Owner>:

Your <date> letter regarding your concern about your neighbor's landscaping arrived today.

[Builder] provides the same landscaping information to all home buyers. Beyond advising home buyers about the proper installation and maintenance of their yards, [Builder] has no authority to enforce the recommendations we provide.

We suggest you approach the neighbor with your concerns. Often a home owner does not consider the impact of grading changes on adjacent yards. If changes to the overall drainage pattern of the area endangers your home's foundation or can be shown to cause water to enter your basement, legal recourse against the neighbor may be available.

[Builder] can provide documentation regarding the overall drainage plan for the community. Finally, your home owners association has strict rules regarding landscaping and plans must be approved by the Design Review Committee. To contact the committee, call <chairperson> at <phone number>. <He/She> can provide you with specific details and current regulations.

If you have any questions, please let me know.

Sincerely,

Warranty Service Manager
[Builder]

This situation is not good for the community, and it is not good for the value of your property. Can we discuss it?" Recognize that you will give in on something to resolve the situation so meet the home owner prepared with options and limits in mind. Work hard to be objective; the company may have made unfair or insensitive decisions that need to be reconsidered. If you learn that the home owner is demanding things you cannot or will not deliver, a conversation with your company attorney might be your last resort.

Everything in the Warranty File

Copies of the contents of a warranty file belong to the company and should be shared only with the current owner of a home. A request for copies of the entire file often comes from a listing real estate agent who may mention "disclosure laws" to get your attention. Disclosure is the obligation of the current seller (the current home owners and their listing agent) and does not apply to you at this point in a resale. Be courteous to all parties as you explain that for reasons of privacy, you will provide the requested information only upon receipt of a written request from the current home owner.

Working with the Media

If you are approached by a reporter, remember that comments made "off the record" are likely to be attributed to "a high-ranking official" or "a source revealed." In addition, reporters might apply a slightly different perspective to a story in an effort to heighten interest in it.

For example, a reporter talked with a builder-developer who was embroiled in a disagreement with a no-growth group. The reporter asked whether engineering for a new community was complete. The developer responded that engineering was in progress and would be completed in one to two weeks. The resulting story stated, "The developer admitted the required engineering had not been done." This statement was just a bit different from what was actually happening, but it made a more intriguing story.

Designated Media Representative Company Spokesperson

Rather than saying something that could be taken out of context to create a different meaning, explain in a friendly and cooperative tone that you "do not have all the facts and do not want to give out incorrect information. The person to speak to is . . . ." and provide the name of your company's designated media representative. The ideal approach is to request a written list of questions and a deadline and then provide written responses.

No "No Comment"

Avoid going to the other extreme by screaming, "No comment," and slamming a door in the face of the reporter. Think about your reaction when a news magazine program attempts to interview the opposing side and someone spits on the camera lens. Your conclusion is likely to be that he or she is guilty as well as rude. Usually if the company can show itself to be a reasonable and caring organization that operates in a reputable manner, the simple fact of a disagreement with a home owner makes a dull story and is likely to get little attention.

Training Activities

The list of situations has no end. However, regular discussion with warranty personnel increases chances of success and decreases errors in judgment.

Procedures and Documents

Review procedures (either online or on paper). Discuss how to proceed during inspections and how to complete checklists appropriately. Provide right-way and wrong-way examples to ensure clarity. Not surprisingly, these discussions often lead to improvements in the procedure or the checklists. They will at least minimize any confusion.

Observations and Mentoring

Have a new warranty staff member observe a veteran, with special attention to the conversations with the home owner. Discuss the results. Reverse the process and have the veteran observe the newcomer; again, discuss the results. Supplement real events with discussion of a case study at staff meetings once a month.

Outside Experts

Bring in outside experts to build knowledge and keep up with new information. Manufacturers often provide trainers or at least training videos. Keep in mind that other departments (certainly construction but also sales, design, and purchasing) might be interested in participating. These broader perspectives can generate lively discussions.

Builder Association Programs and Conferences

Take advantage of educational programs your local home builders association may offer as well as regional, national, and international conferences, such as the International Builders' Show. The rich selection of educational programs currently available to builders is unprecedented. Likewise, review the materials available for purchase from www.builderbooks.com®. You will find literature that can be useful for training and reference. Subscribe to a variety of industry journals and routinely assign reading of the most relevant articles to your staff.

Scopes of Work

Keep personnel updated regarding scopes of work by asking purchasing staff to sit in on a warranty staff meeting once a quarter. Even better, rotate warranty personnel through an internship in purchasing so they understand the processes and results.

Closure

Follow-through brings each warranty item to closure. Closure means that the company, the home owner, and the warranty file all agree that the builder has responded to the item reported. Note that you can achieve closure even when the home owner is less than pleased with your response; closure and satisfaction are not always the same. Simply put, if the home owner reports 14 items, you need to document 14 responses—whether those responses are yes or no.

If your response was yes, your task is to get the work completed and documented. If your response was no, you must explain why, offer information if possible, and document the answer. Chapter 9 discusses such denials in detail. This chapter addresses getting work orders completed and following up with home owners.

Response Time

Traditionally, builders aimed for completing repairs on warranty items on a 30-day time frame. As the focus on customer satisfaction has increased and other businesses have accomplished faster response times, builders have become more aggressive with their deadlines for this aspect of warranty service. Ten workdays is now the goal of choice. You achieve success when you stay within that time frame with 90% of warranty repairs. Weather delays of back-ordered parts can delay some repairs or installation of missing items. These situations should be the exceptions, and they necessitate extra communication with the home owner.

Ensure that you are reviewing accurate and objective data on the timeliness of warranty work. Impressions, hunches, or gut feelings that work is completed quickly can be dangerously inaccurate. Get the facts first; then consider whether you need to improve response times.

If your current response times are excessive, focus on this aspect of service immediately. If your warranty workload is simply overwhelming, consider whether you might be in a "service hole" as described in Chapter 11. Notwithstanding a service hole, from time to time any warranty office can get a bit behind. Set specific targets in increments. Make the goals challenging yet achievable. Alert the trades and in-house personnel who are involved and enlist their enthusiasm

Beware of the unanticipated consequence of incentives. One builder promised in-house warranty technicians $10 for every completed and signed work order they turned in. Before long, dozens of such work orders had accumulated and hundreds of bonus dollars were awarded. Startled, the builder compared this dramatic progress with historical records and found that with the new bonus system in place work orders typically each addressed just one item. Previously home owners had reported items in batches, but that practice suddenly changed shortly after the incentive plan was announced.

and support. Creating some friendly competition, offering an incentive, or planning a celebration for achieving the goal can go a long way toward resolving the situation and getting response times back on track.

Completing Approved Warranty Work

If you use the work-date approach or have an administrator set appointments, for the sake of efficiency, your recordkeeping should include the home owner's contact information. If you rely on trades and home owners to contact each other for scheduling, routine monitoring—discussed later in this chapter—will be your main method of checking on work.

Confirm Appointments

Some builders confirm all known repair appointments both with home owners and their trades. All builders should make a point of confirming repair appointments when either the home owner or the trade involved has a track record of forgetting such commitments. The effort to call or e-mail the parties involved a day or two prior to a repair appointment can forestall considerable extra work later.

On-Site Visit

Depending on the nature or amount of work—or in some cases the personalities involved—attending or at least stopping by during the repair appointment is a wise practice. This visit can be especially productive when you use the work-date approach (described in Chapter 6) because the warranty rep can confirm that all the expected trades arrived and take appropriate action if anyone is missing. During such a visit, the warranty rep can expect to answer questions, address additional items, run interference for a trade with demanding home owners, and confirm that the builder's quality standards are met by the trades' work performance.

Overseeing Repair Work

Major repairs that require the effort of multiple trades over many appointments (such as serious leak repairs, structural claims, or widespread drywall work) can necessitate assigning a warranty rep to the home for most of the time that work is ongoing. This arrangement permits the home owners to come and go, ensures steady progress, allows documentation of repair details, and provides the home owner with a comforting level of attention.

Legal Action Pending

When a home owner initiates legal action, many companies mistakenly put all pending warranty work for this home owner on hold until the legal proceedings are resolved. While this point may be irrelevant if the home owner refuses access to the home, complete the approved warranty work that you can to reduce the number of items under discussion in the legal proceedings. Minimally, document every attempt to perform approved repairs.

Routine Monitoring

Frequent and consistent monitoring are two secrets to successful warranty service. Especially if you rely on home owners and your trades to set their own appointments, pay close attention. For example, follow up on work orders *before* they reach their expiration date. At least once a week, note work orders that remain incomplete according to your records and do something about them. If you wait until monthly reports are compiled, circulated, reviewed, and acted upon, the home owners may be picketing before anything improves. Sending an unpleasant note to a trade who has 14 service orders that are 65 days old is taking too little action too late. Monthly reports are no substitute for quick service intervention.

In order to monitor warranty work effectively, you first need to know what warranty work was ordered and when. To meet this need, establish a reliable system and practice self-discipline in using it for all items. Three suggestions for systems follow.

Computerized System

Effective computer support saves searching through stacks or binders by providing you with an accurate list of work orders that require attention on a daily basis. Figure 8.1 offers an example of lists generated by one system. This screen pops up daily and directs follow-up efforts quickly and productively.

Another approach involves computerized summaries of pending warranty work by home owner and trade, including a brief description of the work ordered. These summaries are useful to warranty reps in the field; they can make notes on these reports daily as information becomes available, such as Mr. Jones is on vacation, the Wilsons' work is completed, the Smiths are waiting for a back-ordered part. The warranty administrator uses these notes and returned work orders to update the pending work summary, usually just before the weekly warranty staff meeting.

The number of states with notice of repair or right to repair laws has increased in recent years. Often along with those laws come strict and precise time frames within which builders must comply in their investigative, communication, and repair actions. Such requirements make effective computer support even more valuable. When every work order, phone call, e-mail, fax, or letter—whether to a home owner or a trade—is documented by date and time, builders provide better service and protect themselves from claims of noncompliance.

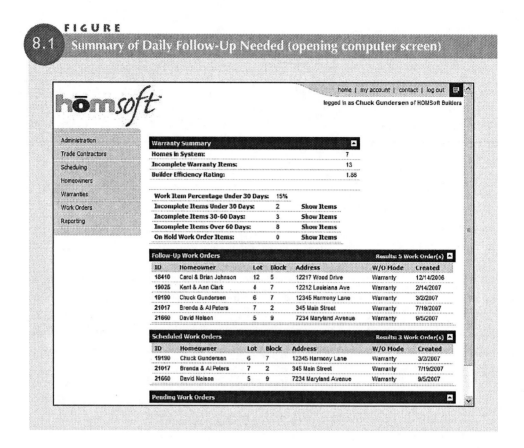

FIGURE

8.1   Summary of Daily Follow-Up Needed (opening computer screen)

Work Order Binder

Your company may still be looking for that perfect software solution for scheduling and tracking warranty activities and your communications with home buyers, trades, suppliers, staff, and others. If so, you can set up an effective manual system as an interim alternative using a three-ring binder as follows:

* Label an index tab for each trade that routinely receives work orders.
* Add a work order log like the one in Figure 8.2 under each tab.
* When you issue a work order, fill in the first four columns of the work order log for that trade.
* Upon completion of the work order, the service technician should sign and date the work order and return it to the warranty office.
* Daily or weekly, depending on volume of work, match the completed work orders with those in each trade's section.
* Note the date of completion on the work-order log and remove the copy of the work order from the binder.
* File the signed copy in the home's warranty file. Dispose of the office binder copy unless it contains relevant notes. If it does, file that copy as well.

FIGURE
8.2 Work Order Log

[Logo]   Work Order Log

Trade contractor _Carl's Cabinets_   Contact _Ryan Mazur_

Phone _555-440-4000_   Cell/pager _555-707-0101_

E-mail _carl@cabinets.com_   Fax _555-440-4003_

Date issued	Completion target	Work order number	Home owner	Completion date	Number of days
6/11	6/21	7394	Buyer	6/18	7
6/14	6/24	7409	Gonzales	6/20	6
7/9	7/19	7435	Patel	7/20	11
7/14	7/24	7482	Chen	7/16	2

What remains in the updated binder is your outstanding work. Beginning at the back of each tabbed section, check for work orders that are approaching their expiration dates. A phone call to the tardy trade and the affected home owner usually explains the delay and gets the process moving again. Make notes directly on the tracking copy of the work order. You can fax the work-order log to your trades as an informal summary of outstanding work orders. They can compare their records and stay on track.

Bulletin Board

Another tracking alternative for a small number of work orders is to put pending work orders on a bulletin board, stacked by trade. This method is often used by small-volume companies in which the superintendent monitors warranty work for a few homes. A quick glance identifies any that require intervention. When work is completed, remove and file the work order. Anyone who comes into the office can readily see the amount of warranty work waiting to be performed.

That visibility can be an advantage or a disadvantage depending on who's looking at the pending work and why. For instance, the plumber who has two "old" service orders may point to an expired work order for the cabinet company and use its slow performance to justify his own (the "everybody's doing it" defense). The opposite effect occurs when a trade sees that her company has the largest stack of work orders and feels pressured to get them completed.

Following Up with Home Owners

Follow-up contact can put a builder's reputation in a class by itself. Think in terms of two types: follow-up with home owners about pending work and follow-up about work you believe is complete. Each can be challenging in its own way; and both are important to successful warranty service.

Secondary Attention for Incomplete Items

In the interest of mental health, expect that some work items will need secondary attention to be completed. Monitor the number of items that fall into that category. The number may be lower than you initially believed. Measuring them as a percentage of overall work provides perspective and a way to track improvement.

Once you know the size of the challenge, analyze the nature of it. If a high volume of service work is involved, breaking it down by trade may be useful. Ask, item by item, "Why was this item not closed?" Categorize the underlying causes and begin taking steps to eliminate as many as possible. Look at three factors for the causes of delays: your office (yes, *your* office), your trades, and your home owners. In the following paragraphs, examples illustrate each category.

Warranty Office

Before accusing others of poor performance, make it a habit to check your own accomplishments.

Administrative Time. Working with the trades and in-house technicians to achieve fast response times will fail to satisfy home owners if their requests linger on a desk for a week before any action occurs. When the warranty office reacts immediately to requests and inspection reports, it reinforces the sense of urgency. Issuing work orders within one day of the inspection should be your goal. Wireless systems permit notice to the trades at the push of a button, either directly from the home at the end of the inspection or from the office at the end of the day.

Complete, Correct Work Orders. The work order that contains complete details about needed repair work is the one most likely to get done. Well-trained warranty reps recognize who needs to be sent to the home and the sequence of repair steps. They anticipate the details that the trade or technician will need to get the work done—color, size, style, and so on.

Access Policy. Warranty work will take longer if you do not accept keys.

Trade Support. Sometimes a service technician will have a question or may need help arranging access to a home or preventing home owner interference with work. Enthusiastic support at such times keeps repair work flowing.

"Who's Going to Pay for This?" Builders sometimes pay for work performed during the warranty period. This payment occurs because companies provide courtesy repairs such as attention to drywall shrinkage damage. A builder also might pay for work after replacing a trade. Enforcing the former trade's repair responsibility can be tough and time consuming, and all the while the home owner is waiting. The new trade contractor will likely expect to be paid for correcting someone else's mistake. Understanding how payments are structured in the trades' contracts and what is included in their scopes of work is vital to appropriate decisions in such matters. If you expect to pay for work, obtain a cost estimate and issue a purchase order or other authorization your company uses.

Legitimate Double Visits and Multiple-Trade Repairs. Some repairs require more than one visit by a trade or more than one trade, each working in a set order. For instance, repair of a plumbing leak is followed by drywall repairs and then by painting. The first repair usually occurs promptly. Steps two and three often drag on. The best approach is to discuss the steps and set up all appointments in one conversation with the home owner.

Was the Answer No or Maybe? Failure to follow through in writing when you deny warranty items can leave the matter unresolved in the home owner's mind. Today's home owners are unlikely to forget any of the items they reported. Often the issue that comes across your desk a second time is the one you failed to respond to clearly or failed to document.

Wrong Trade. Disorganized records or careless research sometimes result in a work order going to the wrong trade. Unfortunately, such work orders could sit on a trade's desk for days or even weeks before a builder learns about the mistake. Train your trades to contact your office immediately if they receive a work order that should have gone elsewhere.

Trades

Trade contractors are likely to live up to your expectations. Make warranty service important in conversations and documents from the beginning, perform consistent follow-up as outlined here, and enlist the support of management and the appropriate staff members when problems occur.

Diluted Urgency. Home owners care more about their warranty items than anyone else. Their homes are the most important homes; their lists are the most important lists to them. The warranty office handles warranty items all day, every day. Every repair item can't be the most important one. This slightly diluted enthusiasm is passed along to a service technician, usually on a work order. The urgency of that item diminishes dramatically because the technician has not heard the home owner's voice, seen the home owner's face, or observed the actual complaint. What is a heartbreaking flaw to a home owner can become routine work to the warranty office, just another nagging detail to the trade contractor, or merely job security to the service technician.

When you routinely enforce time frames, the trades will replace, "If they really wanted it done, someone would have called us," with "Let's get this scheduled. You know we'll hear from [Builder] if we don't." The shift in attitude improves service by restoring some of the original urgency.

Production Pressures. In many companies, the superintendents have more influence over re-contracting with the trades than the warranty staff does. Thus, faced with a choice of satisfying a superintendent or the warranty department, most trades respond to the superintendent. However, when management makes clear that feedback from both production and warranty staff members influence trade-contracting decisions, priorities become more balanced.

Missed Appointments. Whether as a result of a temporary memory lapse, an emergency, or heavy traffic, anyone can miss an appointment occasionally. However, when the behavior is chronic, you might get improvement from being more involved in setting and confirming appointments until new habits take hold, or by sending another trade and back-charging the original company.

Home Owners

Many scheduling challenges stem from the fact that most home owners are extremely busy. Only a few turn setting up an appointment into a power struggle. Sometimes, however, a belligerent home owner refuses to agree to an appointment during established hours or sets up an appointment and then stands up the technician. Occasionally, a home owner harasses the service technician about repair method or quality, or the repairperson and the home owner may have a personality clash. Keep in mind that your goal is the same regardless: Respond to reported items.

Temporarily Unavailable. If your home owner is about to go on vacation, have relatives visit for a month, just started a demanding new job, or for any other reason prefers not to have warranty repairs performed right now, put the work orders on hold. This step is accepted practice under such circumstances.

However, don't rely on home owners to reactivate these work orders. The result usually will be an angry call from the home owners six weeks later, asking when this work will be completed. To avoid this situation, note on your calendar when the hold should end. Check with your home owners on that date to see if they are ready to set up repair appointments. If the warranty period is nearing expiration, limit the hold to between 10 and 14 days. Document all efforts to arrange a repair appointment. Again, what you can prove counts more than what you remember. The letter in Figure 8.3 can help.

Home Owner Missed Appointments. Missed appointments are an acknowledged yet annoying aspect of warranty work. While technicians attempt to keep appointments with home owners who somehow cannot remember to be home when they said they would be, other home owners who would keep their appointments are waiting for service. If a home owner misses three or more scheduled appointments and consistently stands up the service technician, you might consider sending a check for the cost of the work. The letter in Figure 8.4 could accompany the check. You should only use this undesirable solution as a last resort. The defect may remain in the home for others to see; or a second owner may resubmit the com-

FIGURE
8.3   **Work on Hold**

Dear <Home Owner>:

This confirms our conversation of <date> regarding warranty work orders for your home.

In response to your <date> warranty service request, we performed an inspection on <date>. As a result, we issued three work orders on your behalf—copies are <enclosed or attached>.

Because scheduling this work is inconvenient for you at the present time, we have placed these work orders on hold.

Please contact us no later than <date> to reactivate these work orders so that we can complete this work before the associated warranty coverage expires.

If you have any questions, please call me.

Sincerely,

Warranty Service Manager
[Builder]

plaint. However, those concerns may be easier to tolerate than delaying service to other waiting home owners.

Method of Repair. Home owners may demand that you use their repair methods or that you replace an item when a repair is the normal approach. While you should certainly listen carefully to the home owner's opinions, be cautious about using unfamiliar repair procedures or those that you know have not worked well in the past. Following the dictates of the home owner may seem to be service oriented, but when problems develop as a result, they will blame you, and any benefits are lost. The letter in Figure 8.5 addresses repair policies.

Quality of Repair. If a home owner says that repair work is unacceptable, the best response is usually, "What is a convenient time for me to look at this with you?" Trying to resolve such issues over the

In extreme situations the pressure to offer more flexible hours can be severe. For instance, such pressure occurs when the home owner took time off and was "stood up" by a service technician or when an unreasonable amount of repair work requires days and days of access to the home. In such cases, you might work with your trades to arrange appointments outside normal hours or assign a trusted warranty rep to house-sit while work is performed. If you want work performed outside of normal business hours because of the amount of work needed, you may need to pay extra to a trade for such special service if that trade was not at fault in the situation or did not do the original work.

FIGURE
8.4   Check in Lieu of Repair

Dear <Home Owner>:

Work order number _____ was issued on _____ for warranty work
needed in your home. Several attempts to provide the work listed have been
unsuccessful.

After three appointments have been set without obtaining access to your home to
provide a repair, our administrative policies and the terms of your warranty provide
us with the option to pay you for the work indicated. You can then schedule the
needed repair at your convenience and with a repair person of your choice.

A check (number _____) based on the cost to [Builder] for this work is
enclosed. The work order will be voided and the item recorded as resolved in
your file.

If we can provide you with any information that will assist you in resolving this item
on your own, please feel free to call me.

Sincerely,

Builder
[Building company]

phone is a waste of time. The home owner interprets long-winded discussions as
defensive and becomes more upset. Volunteering to re-examine the item shows
interest without committing to any action. If all service technicians know that
home owner complaints result in an inspection of the work, they are more likely
to pay strict attention to details and do a sound repair in the first place.

Confirming Completions

Upon receipt of a completed work order, in as many cases as possible, contact the
home owners—phone, fax, e-mail, postcard (with first-class stamp), letter, or an
in-person visit—to confirm that the issue was resolved. Especially for significant
issues (no heat, carpet replacement, front door lock not locking, and so on), this
contact is essential. When trades know you make follow-up contact, they are likely
to be more meticulous with their repair work. Even if you "know" work is complete, the contact impresses customers. If a home owner is dissatisfied, the company is better off if you hear about it than if that home owner instead tells friends
or colleagues.

As with many aspects of warranty work, this activity can be broken into several
categories, each with its own particular fine points. Follow-up methods can include

> **FIGURE**
> **8.5** **Method of Repair**
>
> Dear <Home Owner>:
>
> In response to your request, work order number _____ was issued on <date>.
>
> The work order provides for repair of <item>. This repair has been performed in the past with excellent results.
>
> The service technician or trade contractor assigned to complete this work will contact you during the next several days to arrange an appointment during normal service hours (8:00 a.m. to 4:00 p.m., Monday through Friday).
>
> As defined by the terms of your limited warranty, the choice of a method of repair is specifically [Builder]'s. You have the option of taking responsibility for correcting the item at your own expense and using a method of your choice.
>
> If you decide to proceed with another method of repair, simply inform the service technician or trade contractor, and we will cancel the work order.
>
> Please call me if you want to discuss this matter or if you have any concerns about the intended work or our procedures.
>
> Sincerely,
>
>
> Warranty Service Manager
> [Builder]

in-person visits, phone calls, postcards, or e-mails. Use more than one method as some home owners will respond to one and not another. In all cases, be cautious in these communications to avoid implying any lack of trust or confidence in repair personnel, especially when the work order was returned without the home owner's signature.

Unsigned, Completed Work Orders. When returned work orders bear only the technician's signature (for example, for exterior work that did not require the home owner's presence), your follow-up is critical. Without this last effort, the home owner may be unaware that the work was performed. Door hangers also address this concern, but your contact draws feedback about *the repair process.* You might also ask whether the home owner has any *remaining questions.* These terms solicit confirmation that the work is done to the customer's satisfaction without suggesting that you lack confidence in your repair people.

Unsigned, Completed, Home Owner Wants Other Repairs. Sometimes work orders are completed in the presence of the home owner, and that owner refuses to sign the work order. If the plumber has performed a repair and requests the home owner's signature, the home owner might say, "I'm not signing anything until I get the fruit tree I was promised in the back yard." Ensure that your repair technicians know that they can note this fact, sign the work order, and return it to warranty. Follow-up in this case begins by asking for feedback on the repair process as described above. No doubt the home owner will mention the desired tree, and the company can then decide what action to take on that issue.

Completed, Signed Work Orders. You can generate considerable goodwill with follow-up contact on items known to be complete. Again, ask for feedback on the process and whether any questions remain. Many warranty reps hesitate to make these calls because they fear that the home owner will respond with "While I've got you on the phone, I've got another list." Be prepared to ask whether any items are urgent. If not, suggest that they be addressed in the year-end visit. If the home owner is uncomfortable with that, set up an appointment to inspect the items. Warranty's goal is to help stimulate referrals. While efficiency is desirable, it is not the primary objective.

Training Activities

Role play various scenarios to ensure that your warranty staff can talk comfortably and in a friendly manner with home owners and your trades about warranty practices comfortably and in a friendly manner. Use the following list for ideas and cull others from your daily experience:

- A technician confirms appointments.
- You make a site visit to check on work in progress.
- A work order is at its expiration date.
- You are certain the work order is complete.
- A work order has been returned, supposedly completed, and you want to confirm.
- A home owner calls to say a trade has missed an appointment for the second time.
- A historically difficult home owner wants an evening appointment for routine work.
- You need to follow up with a home owner on work done by a new trade.
- A trade with a long-term record of less than stellar service turns in a work order—supposedly complete.
- A home owner has missed a third appointment with a trade.
- A work order remains incomplete because details about the repair were omitted.
- A trade insists he needs a purchase order to perform work described on a work order.
- A home owner wants to instruct the technician as to the method of repair.

Denials

Saying no to home owners is often uncomfortable and may even lead to conflict. Still, on occasion, no is the right answer. You can make setting and enforcing boundaries easier and more pleasant by preventing questions to which the answer is no. Volunteering information and communicating in a forthright manner as described in Chapter 1 creates expectations in line with your product and procedures. Ensure as well that you have practiced sound judgment skills as discussed in Chapter 5. Then, when saying no becomes necessary, say it effectively.

Saying No Effectively

When the response to a warranty service request is no, this information is best delivered promptly and in person or at least by phone. Hearing nothing for three weeks and then receiving an impersonal denial of service would upset any home owner.

Similarly, hearing, "I'll check on this and get back to you. . . ." and never hearing another word is equally unsatisfactory. Avoid saying *maybe* and hoping the home owner will forget she or he asked. When you need to investigate, make a firm commitment, "I'll get back to you Friday afternoon." Then do so on time even if all you say is "I'm still working on this issue." No one likes to feel ignored or forgotten, least of all a customer who just bought a new home. Initiating contact will produce much better results than waiting for a frustrated home owner to call you.

Say no in such a way that home owners feel they were treated fairly. Listen to everything they have to say and investigate fully. When you are certain the correct answer is no, as often as possible, express it softly.

Soft No

Whenever you can in your conversations and documents, eliminate the words *no*, *not*, and their many forms including: *can't*, *won't*, *cannot*, and so on. Replace those unfriendly words with the soft no—the no that avoids the word *no*.

Begin by gathering one copy of each document you use with home buyers. Read through every page, red pen in hand, circling every word *no*, *not*, or variation of the two you find. If your paperwork is similar to most companies' paperwork, you will be surprised at how harsh and unfriendly it is.

In addition to updating documents, listen closely to your conversations. When you hear yourself say no, make a mental note of the situation. Later when you have time to think the scene over, look for a new way to say the same thing without using the word *no*. Every time you invent a better phrase, you will expand your hospitable repertoire. When several warranty reps share insights on saying no creatively, they tend to improve their capacity to respond in this way.

Work to reword the information to convey the necessary limits but sound hospitable at the same time. Two simple techniques help. First, ask if you can use another word. Consider words such as avoid, prohibit, exclude, or unavailable.

- Replace "Do *not* expect a dust free surface on your hardwood floors," with "A dust free surface is *impossible* to attain when finishing hardwood floors."
- Replace "Our warranty does *not* cover concrete cracks," with "Our warranty *excludes* concrete cracks."
- Replace "*Don't* plug your food freezer into the GFCI outlet. Our warranty does *not* cover food spoilage," with "*Avoid* plugging your food freezer into a GFCI outlet. Our warranty *excludes* food spoilage that can occur if the freezer trips the GFCI."
  The second technique is to focus on what is available, rather than on what is unavailable.
- Instead of "Our warranty does *not* provide any attention to this item," say, "This item falls under home owner maintenance," or say, "Your personal standards *are even higher* than ours. My job is to make certain *we delivered everything we promised*. If you'd like to make it even better, I'll be happy to *help with information*."
- When a home owner asked why the side basement egress window didn't have a cover like the front one did, the warranty rep responded, "We *don't have to* cover egress windows when they are this far from a door. If you want a cover *you'll have to buy* your own." A more positive response would have been, "We *provide covers when* the egress window is within five feet of a door. If you'd like a cover for this one as well, *you can pick one up* at <nearby store>. I believe they run about $__ and they are easy to install."

Wording can be clear and still friendly. It can make the boundaries easier to accept from the buyers' perspectives, and the information more comfortable to communicate from yours. This communication skill takes practice and thought. The positive results with customers make every bit of effort worthwhile.

Terminal No

The soft no is effective about 75% of the time. For the remaining times, you may need a firmer response. Consider the terminal no for such cases. A terminal no begins with the reasons and uses the word no at the end of the answer. "Because concrete cracks can be caused by . . . , which are out of our control, the warranty excludes cracking. Therefore, the company will not replace your driveway." If you begin your answer with the no, the home owner may not hear your explanation because he or she is thinking of how to change your mind or is simply angry.

No Documentation

The final step in denying a home owner request is to follow up in writing. If you use the warranty service request form shown in Figure 4.3, minor items can be documented quickly by marking the "Maintenance" column on the right side. Having this form printed on NCR paper allows you to give the home owner a copy at the end of the inspection. For more significant items, a conversation followed by a letter is best.

"This letter confirms our conversation regarding . . . ." serves as a useful opening for such letters. This step can forestall the home owner from reporting the item again three months later in hopes of getting a different answer. Put a copy in the warranty file for the home and forward a copy to any trade contractor affected. If the home owner reports the item again, take the approach that, "If something has changed, we'll be glad to reconsider, otherwise our answer remains the same." This response keeps lines of communication open. If something has changed, you should re-inspect and reconsider your response. When a warranty staff person receives a promotion, the person who takes the original position will appreciate the thorough documentation.

Home Owner Maintenance

Home owner requests for service on maintenance items are common in the warranty office. Some of these requests will come from home owners who really don't know what normal maintenance they should perform; others know but believe, "It can't hurt to ask." And some know, but think they can get extras through intimidation. With professional up-front communication, delivery of a complete and clean home, and a warranty processing system that promptly produces results, you will be working from a position of strength. Therefore, denying service on maintenance items will be relatively easy. The letter in Figure 9.1 illustrates one way to confirm such denials. (Paragraph 2 shows a terminal no.)

Home Owner Damage

Denying service on home owner damage to the home can be more difficult. Keep in mind the value of back door standards (see "Cosmetic Damage" in Chapter 5). When arguments for and against the home owner are about even, remember the customer wins all ties. From time to time, you will run out of reasons to bend the rules. When this occurs, use the letter in Figure 9.2 and tailor your denial accordingly.

Letters confirming denials are more likely to get written with computer support. Tailoring standardized letters makes composing them faster and easier than starting with a blank page or screen. Maintain a file folder on your computer or a three-ring binder with copies of commonly used letters as well as those you have found effective for unusual situations. However, personalize the letter enough that the home owner receiving it feels heard.

FIGURE

9.1   Denial of Service: Maintenance Items

Dear <Home Owner>:

Your <date> warranty service request listed seven items which <warranty rep> inspected on <date>. We issued the enclosed work orders on <date> as a result. Two items, the concrete cracks you noted in item 4 and the drywall cracks you noted in number 7, remain to be addressed.

According to the terms of our agreement with you, concrete cracks that exceed a width or vertical displacement of $3/16$ inch qualify for warranty action (filling or patching). The crack in your garage floor is located in the control joint and is less than $1/8$ inch. The crack in your patio is hairline. Neither crack has measurable displacement. Because these cracks are within the agreed upon standards, no action is required.

You can review this information on page <page number> of your [Builder] Home Owner Guide. Attention to these cracks falls under home maintenance, and therefore, they are your responsibility.

The drywall damage (two hairline cracks in the family room, a corner bead in the nook, and three nail pops in the sunroom) are typical of those we commonly see as new homes settle. As a courtesy, we will have the painter caulk and touch up the paint. The new paint will be visible; an exact match is unlikely. You can review this information on page <page number> of your home owner guide.

Your copy of the work order to <painter> is enclosed.

If you have any questions, please contact me.

Sincerely,

Warranty Service Manager
[Builder]

Grading Altered

Of all the things a home owner might do to damage a home, changing the grade is potentially one of the most serious. From wet basements to structural movement to problems with neighbors, the results can be expensive and difficult to correct. Wise builders are able to prove the grades that existed when the home was delivered. The value of this documentation shows clearly in the letter shown in Figure 9.3.

Expired Warranty

The conditions under which you should correct items after the warranty has expired were discussed in Chapter 4. When your evaluation concludes that the items are

9.2   Denial of Service: Home Owner Damage

Dear <Home Owner>:

This letter confirms our conversation regarding your <date> warranty service request.

The items listed—a carpet stain in the family room and a one-inch-long cut in the vinyl in the laundry room—are examples of the cosmetic damage referred to on your orientation form. One of the purposes of the orientation is to confirm that the home is in an acceptable condition. Your orientation forms show your agreement that the home was in acceptable condition.

After the orientation, this type of damage is specifically excluded from warranty coverage. These items fall into the category of home owner maintenance. [Builder] can assist you with information; repairs are your responsibility.

The flooring contractor—<name, phone>—who did the original work on your home can assist you with both of these items. They may be able to clean the spot from the carpet, or if that proves unsuccessful, they can patch it.

Regarding the laundry room vinyl, a patch is possible or you can have the entire floor covering replaced.

Payment for correction of either item will be your responsibility.

If you have any questions, please contact me.

Sincerely,

Warranty Service Manager
[Builder]

cc: <Flooring contractor>

home maintenance, the home owner must be told promptly and courteously. Figure 9.4 shows such a response.

Outside the Scope of the Limited Warranty

In the minds of some home owners, the builder's limited warranty responsibility extends to every problem or worry the home owner has. This attitude can lead to requests for service completely outside the scope of the warranty. Builder limited warranties cover the home as delivered to the home buyer and exclude adjacent land, wildlife, insects, and myriad other items with which home owners may want help. When a request is impossible to fulfill, confirm your conversation with a letter modeled after the one in Figure 9.5.

FIGURE
9.3 **Denial of Service: Home Owner Changes to Grading**

Dear <Home Owner>:

In response to your report of drainage concerns in your side yard, we reviewed the grading certificate for your property and had the surveyors recheck the grades. You will find copies of both readings <attached or enclosed>.

This review shows that the drainage swale on the east side of your home has been changed significantly. Often this type of alteration results from spreading the soil from holes dug for shrubs or trees over adjacent areas. This situation may be what occurred to change the drainage in your yard.

Two choices are available to you to restore proper drainage. One is to re-create the original swale. The second is to install a French drain. Please refer to page <page number> of your [Builder] Home Owner Guide for detailed information.

Either action would be your responsibility. You may have recourse with the landscaping company that did this work.

If you have any questions, please contact me.

Sincerely,

Warranty Service Manager
[Builder]

Difference in Standards

Sometimes a home owner wants quality higher than what your company promised or perhaps higher than what is physically possible for any builder at any price. Such discussions frequently involve subjective or cosmetic aspects of the home for which no measurable standards exist.

In conversations with home owners over such topics, take the approach, "Once again, your standards are even higher than ours. My job is to make certain we delivered everything we sold you. If you want to make it even better, I will be happy to assist with whatever information I have available." Figure 9.6 shows a sample confirming letter.

Referral to Warranty Insurance

If an insurance-backed warranty is part of the coverage provided to home owners, occasionally you may need to refer them to the standards and provisions of that policy. The letter in Figure 9.7 offers a way to respond to such a request. Rather

FIGURE

9.4 **Denial of Service: Warranty Coverage Has Expired**

Dear <Home Owner>:

This letter confirms our conversation on <day, month, date, and year> regarding your <date> warranty service request.

The materials and workmanship coverage provided by [Builder] limited warranty expired on <date>. Therefore, the items you listed are now home maintenance tasks.

For specific information on how to proceed, please refer to the "Caring for Your Home" section, page <page number>of the [Builder] Home Owner Guide. As we discussed, the components of your home are listed alphabetically, and the entries include guidelines on maintenance.

After reviewing the information in the guide, if you have additional questions, please contact me. I will be happy to discuss these maintenance tasks with you.

Sincerely,

Warranty Service Manager
[Builder]

FIGURE

9.5 **Denial of Service: Items Outside the Scope of the Warranty**

Dear <Home Owner>:

This letter confirms the conclusion of our <date> meeting regarding your warranty service request dated <date>, asking that [Builder] remove the snakes from the open space that runs along the back edge of your lot.

As we discussed, the [Builder] limited warranty covers only the home you pur-chased from us and does not extend to the open space behind your property.

Further, the limited warranty specifically excludes insect and animal activity. We have asked the state wildlife division at 555-555-5555 to mail you information regarding snakes and snake safety.

I hope this information will be helpful. If you have any questions, please contact me.

Sincerely,

Warranty Service Manager
[Builder]

FIGURE
9.6   Denial of Service: Items Within Builder's Promised Standards

Dear <Home Owner>:

In response to your <date> warranty service request, we conducted an inspection of the items you listed. We have issued work orders on four items—copies of those work orders are enclosed or attached.

As we discussed during the <date> inspection of the items on your list, the hardwood floor finish in the kitchen and nook areas is showing the normal effects of use and requires no warranty attention. Please refer to your [Builder] Home Owner Guide and to the manufacturer's brochure you received at your orientation for complete details on caring for this flooring.

I regret the misunderstanding we have experienced over this matter. Our literature is clear and forthright. While I recognize that some of your personal standards exceed the standards described in that material, these standards were the terms of our agreement with you at the time our contract was signed. Our intention is to fulfill our warranty obligations as defined by these documents <enclosed or attached>.

If you have any questions, please contact me.

Sincerely,

Warranty Service Manager
[Builder]

than engage in endless arguments, you may get closure faster and be treated more objectively by a third party from the insurance company.

Refer to Attorney

Some home owners may believe they have legitimate claims that have gone unanswered; others may be threatening you based on real or imagined issues in order to gain a financial settlement. Either way, your best defense is to seek legal counsel earlier rather than later. When a home owner has his or her attorney write to you, the most appropriate action is usually to have your attorney respond. Unless you have formal legal training, your conversation or correspondence with a home owner's attorney can easily make a bad situation worse.

Write a rough draft containing the main points you believe need to be covered in the response and ask your attorney to compose and send the final letter. If your documents include a provision for alternative dispute resolution (ADR) such as arbitration, and depending on the circumstances, your attorney's response may point this out.

> **FIGURE**
> **9.7   Referral of Home Owner to Warranty Insurance Policy**
>
> Dear <Home Owner>:
>
> As part of your new home purchase, [Builder] provided an insurance-backed warranty policy. You received information about this policy as part of your purchase agreement, again at the closing on your home purchase, and in the mail from <insurer> shortly after the closing.
>
> That policy includes standards addressing <component or repair in dispute>. A copy of the standards booklet for this coverage is enclosed for your convenience.
>
> While we regret you are dissatisfied with the repair we provided, we believe we have fulfilled the terms of our agreement with you.
>
> If after reviewing the terms of these agreements, you still believe our attention to <item> fails to meet the standards listed, you may want to contact <warranty insurer> to file a claim. In addition to specifying standards and repairs, <warranty insurer> provides for arbitration in the event of a dispute.
>
> If you have any questions, please contact me.
>
> Sincerely,
>
>
> Warranty Service Manager
> [Builder]

As a courtesy and to forestall any accusation that you ignored the home owner's attorney, advise the home owner that you have turned the matter over to your attorney. Your letter should avoid speculation, character analysis, or evaluation of the claim in dispute. Figure 9.8 shows a matter-of-fact way to accomplish this goal.

Angry Letters

Saying no to a home owner can result in the warranty office receiving a stinging rebuke in the form of an editorializing letter listing the many affronts the company has committed against the home owner. Should this occur, keep these suggestions in mind.

- Read the letter, withholding comments. You may think whatever you want, but say nothing.
- Make a copy to work from; file the original with no notes on it.
- Put the copy of the letter aside until the next work day. Allow your natural (and sometimes appropriate) defensiveness, frustration, and anger to subside.

FIGURE
9.8   Referral of a Claim to Builder's Attorney

Dear <Home Owner>:

The letter from your attorney, <name>, dated <date>, arrived today.

We have forwarded this letter and copies of our agreements with you to <builder's attorney> for response.

<He or She> will handle all future communication and will be in touch with your attorney in the near future.

Sincerely,

<President, CEO, or Other Appropriate Officer>
[Building Company]

cc: <Builder's attorney>

- The next day, reread the letter and identify each issue, sifting out the home owner's emotional comments.
- Categorize and prioritize the issues.
- Check the file for background information on each issue. Avoid the trap of thinking you remember everything. Go to original documents.
- Talk to others who may add to what you learn from the file.
- If necessary or appropriate, set up an appointment with the home owner to inspect the items.
- Come to conclusions based on the facts you gather and fair reconsideration of all circumstances.
- Follow through with appropriate actions. Issue work orders or send a letter confirming denial of requested service.
- Check the big picture. Did any of the company systems let the home owner down? Company failures may create pressure to do extras in some situations. If this happens, improve company systems, as needed. Candidates include product design, expectations set from the beginning, purchasing decisions, construction scheduling and supervision, trade performance, staff training, and so on.

Arbitration

Another avenue angry home owners may take is demanding that a neutral third party review their claims. In the regrettable event that a warranty disagreement cannot be resolved between your company and the home owner, having an arbitration or other alternative dispute resolution clause as part of your warranty document can forestall litigation and save everyone time and money.

To use arbitration effectively to settle a dispute, be prepared to present your company's side of the issue.

* Organize the file in chronological order.
* Read all of it.
* Ask what you can *prove*. (This proof is one reason to do all of the documentation discussed earlier.)
* Obtain and document historical background from other personnel.
* Confirm that all work orders issued are accounted for, and that you know their status.
* Add copies of applicable manufacturers' warranties to the file.
* Alert involved personnel and trades; confirm the arbitration date and their availability to attend.
* Meet with these participants well in advance to discuss the events that led to the disagreement and anticipate points the opposing side might raise in arbitration.
* Arrive early and well-rested for the arbitration session.

Predictable Situations

Regardless of the circumstances of any meeting, conversation, or call with a home owner, keep in mind that courtesy is always in order, even if the repair requested is not. Some home owners do make excessive demands; you may even find yourself wondering about the home owner's integrity or mental health. Enforcing a denial may require considerable effort when you face a persistent or intimidating personality.

Abusive Home Owner

No employee should be expected to tolerate verbal abuse, whether in the form of threats, foul language, or other intimidation techniques. Calmly end such conversations by saying the following:

"Mrs. Jones, I understand that you are upset. I'd like to work with you to find a solution. However, I am going to put some boundaries on how we communicate with each other. If you will stop using that kind of language, I'll continue to talk with you. Otherwise, I will end this conversation and call you tomorrow to discuss this further."

Your job is to make fair decisions based on facts and circumstances, not to second guess the home owner's motivation or try to improve the customer's personality (however tempting that might be). Saying no to a customer is okay. Being rude about it is never okay.

If the abusive behavior continues, calmly leave the meeting or gently hang up the phone. As soon as possible complete an incident report like the one shown in Figure 9.9. Mark your calendar and be certain to place the promised call the next day.

Baiting

When a home owner baits you, he or she uses your own words or actions to trap you into giving the answer the home owner wants to hear. For example, "You people advertise that customer satisfaction is so important, let's see you back it up." Two things can be true. Use the word *and* rather than *but*. In this case a good response might be "Yes, customer satisfaction is a priority for our company, *and* this item is a home maintenance issue."

Dishonest Home Owner

Rarely, but possibly, a home owner may show him- or herself to be less than trustworthy. A small percentage of home owners actually watch for opportunities to coerce you into providing extras. When you suspect that you are working with such a character, take the following precautions:

* Schedule two people from your company for all meetings, ideally a man and a woman. Each gender will recall different details. Together they offer a more complete view of what occurred, and you have two witnesses.
* Document every issue discussed, the decision made on each issue, and any time-frame commitment.
* Take time to think through unusual issues or questions the home owner raises. "That's an important issue. Let me investigate the details, and I'll get back to you <date and time>."
* Be cautious about changing well-established policies or methods. Problems often develop when builders venture into unfamiliar territory.
* Follow up promptly and in writing with a well-researched and carefully worded summary. These summaries should be technically correct, legally appropriate, and diplomatically—but clearly—phrased.
* Establish a relationship with an attorney experienced in residential construction and contract issues. Contact this person sooner rather than later to discuss the situation and obtain guidance.
* Network with other professionals. Knowing you are not the only company dealing with dishonest customers can be comforting, and an exchange of ideas can be beneficial.
* Avoid allowing an exceptionally difficult home owner to make you cynical. Maintain a healthy perspective; most home owners are fair-minded, reasonable people who just want a new home.
* Learn from your experiences: Fine-tune paperwork, streamline processing systems, and improve your skills.

FIGURE
9.9 **Incident Report**

Incident Report

Date 8/13/20–	Time 9:15 am

Location 678 West Port Lane	

Participants	Witnesses
Mrs. Matthews	Roger Mazur, Carl's Cabinets

Events

During a routine warranty inspection requested by Mrs. Matthews, Roger arrived to install a cabinet door that had been ordered 4 weeks ago. (reference work order #7498)

Mrs. Matthews said the new door was unacceptable, that it did not match the others in her kitchen.

Roger attempted to explain about wood grain and stain variations and Mrs. Matthews began to cry and yelled at both of us to get "out of her damned house or else."

Roger left, taking the new door with him.

I attempted to discuss an alternative plan with Mrs. Matthews, but she refused to talk to me and pointed at the door.

I departed at that point.

The cabinet issue remains unresolved. Be aware that the original work order to replace the door was based on a courtesy repair as the home owner has been in the home over a year and the door was damaged in normal use. No previous mention of problems with cabinetry appears in the warranty file.

Further complication is that I did not complete the warranty inspection that I was there to conduct.

Signature Harry Beasley	Date 8/14/20–

Going Over Your Head

When a home buyer does not like your answer, he or she may appeal to a higher authority in the hope of obtaining a different response. Resist taking this behavior personally. The home owner would take this step even if you and the higher authority traded places. The solution is training the higher authority to respond in a way that preserves your credibility while allowing the home owner an appeal process. The higher authority should take the following steps:

* Listen to the home owner and take notes.
* Thank the home owner for bringing the situation to the company's attention.
* Commit to a review of all facts and circumstances.
* Promise an update within a stated time frame (one to three business days in most cases).
* Contact you to discuss the situation and consider these questions:
  – Did you let this situation get personal? Often with a difficult home owner, the tendency to go "by the book" is strong and can get in the way of common sense exceptions.
  – Do external circumstances make keeping peaceful relations essential? If your company is about to request rezoning of a parcel of land nearby, the last thing it needs is an angry home owner picketing the site.
* Involve you in any meetings or inspections with the home owner (unless doing so would escalate the situation still further).
* If the final decision is to repeat the denial, the higher authority should deliver the answer.
* If the final decision is to change the answer to a yes, you should deliver that answer and oversee the adjustment. "Mrs. Jones, as you requested, we reviewed your situation. After thinking about the points you made, I saw the validity of many of them, and I was able to convince <higher authority> of the fairness of taking care of this for you."

"I Don't Care What Your Guide Says"

If your colleagues fail to instill respect for the authority of your home owner guide in home buyers, you will face this comment regularly. Even if your colleagues do their part, you will still occasionally face this remark. Some home owner's psychological make up prevents them from accepting that anyone else has any authority. Techniques under "Difference in Standards" might help. Some home owner's expectations exceed all reasonable interpretations of your company's commitments. In such cases, your choices are limited to giving in to unfair demands or having a disgruntled home owner. Pleasing 100% of your home owners 100% of the time is extremely unlikely.

"I Paid $_____ for My Home"

Sometimes a home buyer uses the price of the home to justify a demand. Avoid statements such as "Well, you didn't buy a custom home you know!" Or worse,

"What'd you expect? This is our cheapest floor plan." Instead say, "At half that price, we owe you what we promised. Please show me what you're concerned about."

You might ask, "What was it in our communication with you that made you expect . . . ?" If the home owner can point to a document that specifies the item in question, or if a reasonable person could have misunderstood what the company promised, you should probably provide a correction. Then correct whatever caused the confusion to prevent a recurrence.

"I Used to Be in Construction"

A strong response to this claim is "I'm glad to hear that. So many of our buyers don't understand that builders have many right ways to build a home. Where was it that you worked in construction?" Lead the conversation around to "We do it this way here."

"I've Already Talked to My Attorney"

Nearly everyone has an attorney or at least knows this threat can produce results. Review the facts and documents. If you believe you are on firm ground, keep lines of communication open and at the same time show that the threat will not change your response. "I regret that you feel that was necessary. If you think I've over-looked something, I'll be happy to review the details again. Please be sure your attorney has copies of your contract and the warranty documents. We intend to fulfill the terms of both."

"I Want to Talk to the Owner"

Respond calmly and courteously. "You are always welcome to do so. One thing I'd point out to you, however, is that I am working within the guidelines the owner has given me." Whether you follow this conversation by alerting your company owner is a matter for you to discuss with the owner. Some owners prefer advanced notice; others want to be able to say with complete honesty, "I appreciate your bringing the matter to our attention, and we'll reconsider your situation. We'll get back to you by Thursday." Refer to "Going Over Your Head" for more suggestions.

Sales Office Visit

Occasionally an angry home owner wants to create a scene in front of several prospects. For the prospect, the interesting aspect of this scene is how the salesperson handles it. When the salesperson stays calm and proceeds with a clear sense of direction, the salesperson shows confidence in the effectiveness of the warranty system. The salesperson should offer the home owner a seat, a refreshment, and take down the details of the request, either on a standard service request

This point is important enough to repeat: If you receive correspondence from a home owner's attorney, seek help from your company's attorney. Avoid carrying on conversations or correspondence with an attorney without guidance from your attorney. You can inadvertently say something that compromises your company's otherwise legitimate defense.

form for faxing to warranty or by directly inputting them into the computer for e-mailing to warranty.

Training

Denying service while retaining customer goodwill is one of the most difficult warranty skills to master. Practice has no substitute. The more attention and effort applied, the better your skills will become.

In addition to role playing the various scenarios, also report on a negative customer service incident in which you were the customer and a company said no to something you requested. Describe the situation and your reaction to how it was handled; discuss how it could have been handled better.

Trade Contractors

Every home owner contact with a trade contractor contributes to the home owner's impression of the builder. Home owners apply the same criteria to a trade's performance as they apply to the builder's staff: response time, attitude, and ability. The potential impact of trades on customer satisfaction—and future business—is too important to leave to chance.

Be the Builder to Work For

Builders often hold their heads over the question of how to motivate trade contractors to respond promptly to warranty work orders. While withholding payments definitely gets the attention of an unresponsive trade, this action comes with unpleasant side effects. Besides angering those affected, it may constitute a breach of contract as well. By the time this step has any impact, your home owner has experienced poor service. Use this negative technique only as a last resort. Instead, approach the issue from a more positive perspective. If you were a trade contractor, which builder would get your best prices, your best technicians, and your best service?

Screen Warranty Items

Avoid sending trade A to correct what trade B should fix. Warranty reps need to be able to diagnose accurately what attention is needed and who should provide it.

Provide Complete and Accurate Information

Incorrect or incomplete descriptions of the warranty work needed are extremely frustrating for tradespeople, and they do nothing to improve the service to home owners. Besides correct addresses and phone numbers, provide complete instructions—including color, sizes, and so on—needed for one-stop repairs.

Much of the quality of the service your home owners receive reflects the trade contractors' attitudes toward your company. Make working for your company so appealing that trade contractors are driven to protect your interests in the hope of continuing to work for you.

Enforce Boundaries

Protection from excessive home owner demands means the builder stands between the home owner and the trade contractor. Sending trade contractors to perform repairs or service on nonwarranty items for home owners is unfair unless you pay for the extra attention.

Establish Lines of Communication

Some companies have found that, when a community opens, an on-site meeting to review key operational details is productive. Support this approach with weekly or biweekly meetings, again on-site, to make announcements, review safety, schedule work, handle service matters, and answer any questions. These meetings can provide a forum for trade-to-trade communication as well.

Maintain Two-Way Communication

Ask for feedback about your procedures and provide feedback to your trade contractors about their performances. Discuss performance and quality issues openly and often.

Show Appreciation

Keep in mind the value of a thank you for a job well done. This thank you might include simple gestures such as taking someone to lunch, a phone call, a letter from the company owner, or recognition in the your company newsletter. Send the trade contractors copies of complimentary letters from home owners that mention them.

Get the Facts

When a home owner complains about a trade's performance, ask to hear the trade contractor's version of the story or event. This practice is not only fair, but it can prevent you from having to apologize for making accusations before you have all the facts.

Pay on Time

Just as superintendents and home owners sometimes wonder when a trade will show up, the trades sometimes wonder when their checks will appear. Many builders have discovered the benefits of paying more often than other companies in their regions. When a problem arises with an invoice, communicate with the trade contractor immediately to work out the problem.

Get to Know Your Trades

Call once in a while or stop at a trade's place of business without asking them to do anything and without complaining. Showing interest in how the trades conduct business builds loyalty as well as providing company personnel with an opportunity to learn technical details from the trade contractor.

Include Trades in Company Training

Where practical, involve trade personnel in company training. Whether technical information or communication skills are addressed, better-trained trades means better service for your home owners.

Plan Social Events

Holidays or just summer can be a good reason for a social gathering, and they work even better if families are invited. The goals are to establish healthy relationships and build loyalty.

Ask the Trades to Train the Warranty Staff

Your in-house personnel often can correct minor trade contractor items. Scheduling trade contractor training sessions on such items for your warranty staff can result in faster service for home owners and fewer interruptions of the trades' work on new homes. In appreciation of this support, most trades are happy to provide small parts, tools, and other repair materials to your personnel. Track the number of such repairs, however, to prevent the trades from relying too heavily on the work of your staff.

Conduct Surveys

Quarterly or semiannually, ask the trades to complete a one-page survey questionnaire about operational details. Likewise, have field and warranty personnel complete a rating form about the trades you work with. Use this feedback to plan improvements. Small companies can conduct this survey informally by phone.

Trade Contractor Liaison

To provide increased attention to trade contractors and to improve communication with them, some companies designate a staff member as the liaison between the trades and the company. Figure 10.1 is a sample job description for this position. The trade contractor liaison takes over after the company and a trade contractor have both signed a contract specifying that the subcontractor is eligible to provide certain work for the builder. Although this procedure was developed with larger companies in mind, some of the specific duties are applicable to small firms where one person should be designated as the main contact for the trades.

Contracting with Trade Contractors

Few builders forget to discuss specifications, scheduling, and prices. They seldom overlook payment terms, and they thoroughly cover on-site supervision and the location of the job. Builders and trade contractors routinely discuss these points because they are important to both the builder and the trades. Adding warranty service to this list raises the topic to the same level of importance.

FIGURE

10.1 **Job Description: Trade Contractor Liaison**

Position Title: Trade Contractor Liaison

General Purpose: Facilitate communication among purchasing, warranty, and field personnel and all trade contractors and suppliers for continuous improvement in customer satisfaction, product, process, and costs.

Duties and Responsibilities:

Trades and Suppliers

- Meet with each newly contracted trade contractor or supplier and conduct thorough orientation into company processes and procedures.
- Coordinate with accounting to ensure required insurance certificates or other documentation is on file and current.
- Survey trades and suppliers for evaluation of company personnel and policies.
- Coordinate recognition and appreciation of exceptional service.
- Implement requests from warranty to hold funds as needed.

Construction Process

- Confirm that field personnel have current copies of scopes of work and manufacturer's installation instructions.
- Work with construction and trades to develop written quality checklists.
- As needed, arrange for community preconstruction meetings with superintendents and other on-site personnel to coordinate with trades and suppliers regarding jobsite issues.
- Ensure that all trades, suppliers, and supervisors have up to date, accurate blueprints.
- Meet monthly or as needed with warranty manager to review responsiveness to warranty work orders and customer treatment.
- Assist construction in resolving confusion between the trades and the company or among the trades.
- Review work in the field with construction superintendents to confirm that trade contractors are meeting the specifications and scopes of work for which they contracted.
- On a quarterly basis, survey field personnel and warranty staff for evaluation of trades and suppliers.
- Organize annual social events, minimally including a summer picnic and a winter party to celebrate successes and demonstrate company appreciation for the trade contractors.

Sales

- Notify sales and design center personnel of significant changes in techniques or materials.

FIGURE
10.1 *Continued*

- Update sales and design center personnel regarding information needed for prompt and accurate response to change requests.
- Coordinate pricing of custom changes to ensure prompt response to home buyers.

Warranty

- Review warranty reports to identify recurring items for elimination.
- Review individual trade and supplier performance of warranty items.

Reports to: Purchasing Agent

Contract and Scope of Work

The contract between the company and a trade contractor should document the trade's warranty obligations including several vital topics. You can find detailed information and sample wording in the previously mentioned publication, *Contracts and Liability*, fifth edition, by David Jaffe and David Crump and published by BuilderBooks.

Warranty Coverage Dates

A year is a year unless the trade contractor starts the warranty year the day it installs equipment, and you start the warranty on the closing date. Then a year becomes an argument. For example, when the contractor says, "The garage door has been on that house for sixteen months," your response may be "Yes, but the home owner has only been in the house for ten months and the door is messed up." An argument over who will pay for the repair follows while the home owner waits for help. For the sake of sanity, builders prefer that all materials and workmanship warranties begin on the closing date.

Emergency Response

Trade contractors should understand your definition of emergency as presented in your home owner guide. If provision of emergency service is appropriate, you should document in your agreements that the trade contractors need to provide 24-hour, 7-day emergency warranty service. (Make a habit of occasionally testing the emergency response system in the evenings or on weekends.)

Nonemergency Response

To save arguments later, make your target response time part of the trade's contract, outline the repercussions of

Document the event that begins the warranty period and its duration in the written agreements with trade contractors. Include any grace period that your home owners have at the end of their warranties.

late warranty work, and emphasize the need for prompt communication in the event of a delay in the work.

Warranty Service Orientation

When your company contracts with a new trade, set up an appointment to meet with that firm's owner and the trade's warranty service person. Create a standard packet for these meetings to assure consistency. Time spent covering this information is an investment in your reputation. A sample agenda suggesting materials to include and topics to discuss appears in Figure 10.2.

Warranty Commitment

Briefly review portions of documents that establish the warranty obligation. Send the new trade a copy of these items prior to the orientation meeting, including a copy of the section of your home owner manual that relates to that trade. Ask for comments and ideas about this material; trade contractors often contribute technical information that is valuable in educating customers.

Procedures and Paperwork

Trades should learn how the builder handles service requests from start to finish. Using your home owner manual as a point of reference, provide the trade with an overview of standard warranty procedures, including all warranty forms and explanations of how to use each of them. If the trade contractors know that a warranty representative has confirmed the legitimacy of the claim, prompt attention is more likely.

Review several typical work orders and, if the trade you are orienting provides emergency service, also review a confirming work order: "Confirming emergency phone call of December 12." Understanding that a confirming work order is merely the paperwork catching up to the emergency already reported helps trades avoid duplication of effort.

Share a recent pending work order report (or go through your computer print out or your binder of outstanding work orders) and discuss how you monitor completions, the repercussions of late work, and the need for the trade to alert you if unavoidable delays occur in completing warranty work. Clarify that a timely response and prompt communication are essential to home owner satisfaction and, therefore, to the long-term business relationship with your company.

Warranty Repair Appointment Guidelines

Your customized version of Warranty Repair Appointment Guidelines (Figure 6.3) is an excellent base for discussion of daily warranty service behavior. Furthermore, by explaining that your in-house warranty personnel follow these same guidelines, you demonstrate a consistent commitment to home owner satisfaction.

FIGURE
10.2 Warranty Service Orientation Agenda

Trade contractor __*Carl's Cabinets*__   Contact __*Roger Mazur*__

Street address __*5697 Olson Blvd. #17-B*__   Phone __*555-440-4000*__

City, state, zip __*Bailey, CO 88888*__   Fax __*555-440-4003*__

E-mail address __*carl@cabinets.com*__

Warranty Commitment	Notes
Trade contract clause regarding warranty	*Email is good for work orders—*

Warranty Commitment

Trade contract clause regarding warranty
- Start date and duration of coverage
- Limited warranty document
- Purchase agreement clause regarding warranty
- Home owner guide entry
  - Maintenance hints
  - Warranty commitment
  - Courtesy extras

Procedures and Paperwork

- Warranty contacts
  - Standard
  - Miscellaneous
  - Emergency
- Warranty Service Request form
- Inspection form
- Work order
- Confirming work order for emergencies
- Pending work report
- Repercussions of late work orders

Warranty Repair Appointment

Guidelines (see attached document)

Trade Contractor Services Procedures

- Paid service work
- Back charges
- Service plan

Notes

Email is good for work orders—

Roger checks from his home office

each morning.

Trade __*Carl Cabinet*__   Date __*5/15/20—*__

Builder __*Harry Beasley*__   Date __*5/15/20—*__

Trade Procedures

As helpful as your standard company documents are in orienting new trades, you also need to discuss in these meetings the issues unique to trade contractors. Include copies of related forms and documents in your trade orientation materials, such as the following items:

- warranty service request
- Warranty Repair Appointment Guidelines
- service plan (see Figure 10.3)
- work order
- confirmation work order for emergencies

Paid Service Work

Knowing that you do not expect the trades to bear the cost of courtesy repairs or repairs needed because of errors made by other trades results in quicker service for the home owner. In emergency situations, a purchase order number given over the phone can instigate repair activity while the paperwork catches up. Be sure to provide the trade contractor contact information in writing for the person in your company who has the authority to issue purchase orders for such work.

Back Charges

Business sense and common sense can be compatible. Nickel-and-dime back charges can backfire when a builder needs some extra help from a trade. If an in-house service technician can repair a minor item that is technically the warranty responsibility of a trade, have the in-house technician or the warranty rep who inspects the problem handle it. Doing so saves you the administrative effort of issuing and tracking another service order, and the home owner avoids another service appointment. Similarly, a wise trade occasionally will do a little something extra and not bill the builder for it. A give-and-take approach promotes a healthy team attitude that can save everyone a lot of hassle.

At some point, back charges will be necessary. A plumbing leak that damages drywall and a hardwood floor involves repair costs that the builder should not have to absorb. Alert the trade and explain the conditions and the amount involved. A telephone call with written follow-up can prevent conflicts. The trade's insurance may cover some of these charges, so the trade will want to initiate the processing of such claims immediately.

Service Plan

Ask each trade to complete a service plan similar to Figure 10.3. This form gathers standard information from each trade and documents any special procedures a builder needs to know. Whether the information goes into a database that all staff can access or in hard copies of the forms for everyone, the information can make reaching a trade easier.

FIGURE

10.3 Service Plan

Date _1/12/20–_

Company	Carl's Cabinets
Nature of work	Cabinets
Unit number	17-B
Street address	5697 Olson Blvd.
City, state, zip	Bailey, CO 88888
Phone	555-440-4000
Fax	555-440-4003
E-mail	carl@cabinets.com
Mobile	555-702-0101
Pager	555-707-0101
Service manager	Roger Mazur
Business hours	7:00 am–5:00 pm
Service hours	8:00 am–4:00 pm
Best time to call	7:00–8:00 am
Emergency	NA
Work orders	email via system
Notes	Roger checks e-mail each morning from home office and can go directly to first home if necessary. Can do first appt. at 7:00 if critical but prefers not to do so regularly. Parts generally take 4–6 weeks to obtain.

> A builder and a trade each have a big investment in their relationship. A sincere effort to improve communication and service is worth a try and may resolve a problem.

Difficult Trades

If one or more trade contractors are out of step with the rest of the team, you have three options: (a) solve the problems with the trade, (b) replace the trade, or (c) continue putting up with unsatisfactory service. This last choice is unacceptable to a company concerned about its reputation. However, repeated turnover of trade contractors is time-consuming and causes confusion in home owner services.

Set aside the assumption that the fault lies totally with the trade. Schedule a productive problem-solving session with the trade company owner and the trade's service person. Prepare for such a meeting by reviewing the agenda used to orient new trades about your company's service expectations, and note the areas that need improvement. Compile specific examples of poor service events so the trade knows exactly what needs to change.

By approaching the meeting in a positive way, you can cover the major points without alienating the trade. To show goodwill, approach the discussion by acknowledging some responsibility: "Perhaps we did not communicate our service expectations clearly enough," or "We've come to realize that as a company, we must all provide better service, and we need your help to do that."

Give the trade an opportunity to explain possible misunderstandings and also to mention areas of your system that may impede good service from this trade. Watch closely for improvement.

When a Trade Disappears

When a trade contractor is no longer available, regardless of the circumstances, your first concern is what arrangements you can make to continue providing warranty service to home owners. Begin by analyzing the potential effects:

* How many home owners are affected?
* How many months of warranty coverage remain for each of the homes?
* What has been the rate of callbacks for this trade?
* How many warranty service orders are currently outstanding?
* What resources are available?
* How much money does your company currently owe the trade?
* Will any former employees of the trade continue the business? If so, are they acceptable candidates to complete the work?

The home owners should receive the same services they would have been entitled to if the trade contractor had continued to work for you. The question is who is going to pay for them? Outgoing trade contractors may have the integrity to meet their warranty obligations, but even with good intentions, results may be limited to their actual resources to do so.

The replacement trade contractor may commit to a limited amount of warranty service as an inducement to your company to contract with the firm. To avoid any misunderstandings, both of you should clarify exactly what the new trade will do and for which homes. Put this information in an agreement signed by both of you. If all else fails, your company simply orders the work and pays the bills. By continuing to provide professional attention as you always have, you will gain more than you will lose. Protect your reputation by standing behind your product even if you do so alone.

When to Replace a Trade Contractor

Some builders seem to enjoy discharging trade contractors. While repeated replacements may be exciting, ultimately the effect is confusion and finger-pointing. No one is ever certain who did which job, what the status of installation or repair work is, or whom to call in a legitimate emergency.

When several attempts to gain a trade's cooperation have failed to produce any improvement, perhaps the trade company is not putting enough effort into improving its performance. Still, consider carefully the impact of dismissing a trade and ask yourself the following questions: How is the trade's performance in other aspects of the relationship such as technical skills, adherence to construction schedules, and pricing? How many homes are involved in the complaint? Changing trades for the last two houses in a subdivision of 116 homes makes little sense. On the other hand, word could get around among other trades that your company tolerates poor service performance. This reputation can result in other trades putting forth less effort and start a downward spiral.

Therefore, contract carefully, orient thoroughly, and communicate regularly to create a strong team of service-oriented trades.

Training Activities

At first glance, what warranty needs to know may seem to be which trade to send to a home to do a repair. However, warranty reps must learn much more than that information to achieve warranty excellence:

* Begin with your basic trade contract and follow with a systematic review of scopes of work. Warranty personnel are unlikely to memorize all the details in these documents but familiarity means they will know where to go to look up something.
* Schedule time for warranty personnel to visit with purchasing to learn more details about how the company contracts with trades.
* Arrange field visits by warranty staff to observe trades working to improve their technical knowledge.
* Invite manufacturers to conduct training sessions for the warranty staff, especially appliance companies and those that specialize in heating, ventilation,

and air-conditioning (HVAC) to build the warranty personnel's technical knowledge and improve their confidence.

* Make visits to trade contractors' places of business, both to learn and to socialize a bit.

* Role play conversations staff might have with trade personnel regarding performance or behavioral errors. Base the discussions on points covered in your Warranty Repair Appointment Guidelines.

* Ask accounting to explain how back charges are documented and what details are needed to successfully collect on a mishap.

* Use a case study approach to practice problem solving with your team. Base the cases on actual events from your experience.

Warranty Reports

The insight that warranty departments can offer is too valuable for builders to overlook. Well-conceived and effectively used warranty reports help increase customer satisfaction and decrease repair costs, but only if these reports are accurate and reviewed regularly. All reports should be precise, relevant, and available to staff people who need the information. In turn, the builder must use these reports to improve policies, processes, and services. If knowledge is power, it is also responsibility.

Benefits

Objective reports help you convince others to make necessary changes. (Yelling does not work well, or for long.) Look for patterns in the data. When builders apply the data appropriately, the following results may occur.

- Product performance influences future design and purchasing decisions.
- Repair data calls attention to inadequate supervision during construction.
- Warranty staff work loads are judged more accurately.
- Problems with the trades or suppliers can be quantified, making resolution easier.
- If buyer expectations conflict with the builder's warranty commitment, this is evident in reports of items denied.

Essential Reports

The nature of warranty work lends itself to many types of reports. Begin with the essentials: number, nature, and completion of warranty items. Certainly costs need to be reviewed routinely as well. Once these basics are mastered, you may want to add other subjects to your tracking routine. Several are suggested in this chapter, and you will think of still more.

Purchasing managers, project managers, and superintendents respond to hard facts. "I have a hunch. . . ." will not produce changes, but this statement might, "We've spent $3,822 performing repairs in 11 homes because of this error."

Number of Warranty Items

Warranty personnel need an "early warning system"—a set of criteria that alert them when warranty work is heading for trouble. The first indicator in this system is the volume of warranty work.

Calculate the number of warranty service requests and total items you would receive if every home owner used your system exactly as you describe it in your home owner guide. For example, if you build 30 homes a year and your procedures include two standard checkpoints, one at 30 or 60 days and another at 10 or 11 months, that system would generate 60 lists per year or an average of 5 lists a month. Add 10 to 15% to this total to allow for emergencies, home owners (or homes) that need extra attention, and reports of items from out-of-warranty home owners. For this example, that calculation would mean 69 lists.

Considering the size and complexity of your product, how many warranty items per list should you expect? While everyone agrees the ideal is zero, reality suggests you will have some. If you believe 6 items per list to be a reasonable average number of items, you should anticipate warranty addressing 414 items per year (69 lists times 6 items) or 36 per month. If your records shows the volume of warranty work is at 976 items per year (more than 80 per month), your early warning system alerts you to look for the causes.

Volume is the most critical item for most builders to monitor. However, your early warning system should include other factors besides volume. Other early warning signs usually also result from excessive volume.

Nature of Warranty Items

Builders frequently describe repair work based on the number of outstanding lists: "We have 21 outstanding lists." Interesting, but insufficient. You cannot tell how much work those lists represent, how much of it has been completed, or anything about the nature of the work needed. Are they minor items? Structural? Cosmetic? Material failure? Workmanship issues? Nor can you decipher how many items are purely warranty items and how many are courtesy repairs.

To make use of warranty repair data, you need item-level details. For example, when a builder says "We have 108 work orders containing 417 items; 9 percent are cabinet items, 11 percent are floor covering," a clearer picture develops. With objective information, you can select appropriate improvement targets, develop effective strategies, and make convincing presentations to those people whose cooperation you need to eliminate recurring items. Reports based on a particular floor plan, individual trades, individual home owners, specific subdivisions, or a range of dates—among other sorts of criteria—should be available. Figure 11.1 offers one example of this type of summary. Share such information with your trades and ask them how they might eliminate recurring items.

Remember to share some of this information (or at least a summary of it) with your salespeople. They can speak to home buyers with more enthusiasm and confidence when they are well-informed about warranty efforts.

FIGURE

11.1 Warranty Completion Report

Warranty Completion Report
(one-year summary)

Community: Shoreline Estates II

Date Range: 03-01-20__ through 03-31-20__

Warranty ID	Total	Warranty	Nonwarranty	Courtesy	% Warr	% NW	% Ctsy
Doors	600	255	2	343	43%	0%	57%
Carpentry	555	350	15	190	63%	3%	34%
Plumbing	519	435	7	77	84%	1%	15%
Painting	458	128	55	275	28%	12%	60%
Drywall	439	219	43	177	50%	10%	40%
Flooring	224	88	22	114	39%	10%	51%
Electrical	214	76	0	138	36%	0%	64%
HVAC	206	105	0	101	51%	0%	49%
Windows	176	95	6	75	54%	3%	43%
Siding	134	78	0	56	58%	0%	42%
Cabinetry	129	45	4	80	35%	3%	62%
Landscaping	116	12	48	56	10%	41%	48%
Concrete	109	23	48	38	21%	44%	35%
Miscellaneous	66	0	0	66	0%	0%	100%
Masonry	51	23	0	28	45%	0%	55%
Roofing	39	8	0	31	21%	0%	79%
Appliances	37	0	0	37	0%	0%	100%
Home Owners Assn	36	5	31	0	14%	86%	0%
Tile	35	23	0	12	66%	0%	34%
Fireplace & chimney	30	14	0	16	47%	0%	53%
Leaks: Water or air	27	27	0	0	100%	0%	0%
Glass & mirrors	23	4	0	19	17%	0%	83%
Guttering	21	14	0	7	67%	0%	33%
Countertops	19	12	0	7	63%	0%	37%
Driveway	16	0	7	9	0%	44%	56%
Garage door	16	16	0	0	100%	0%	0%
Insulation	15	15	0	0	100%	0%	0%
Sprinklers	10	10	0	0	100%	0%	0%
Site preparation	8	3	0	5	38%	0%	63%
Sewer/Water	5	5	0	0	100%	0%	0%
Cleaning	2	0	0	2	0%	0%	100%
Asphalt	1	0	0	1	0%	0%	100%
Decorating	1	0	0	1	0%	0%	100%
Stucco	1	1	0	0	100%	0%	0%
Water treatment	1	1	0	0	100%	0%	0%
Millwork	1	0	0	1	0%	0%	100%

*D-days, H-hours, M-minutes.

Source: HOMsoft™, Burnsville, Minnesota

The work order to install a replacement cabinet knob may seem like a small detail when it is mixed with dozens of other warranty items on your desk. However, to the home owner who is waiting for this item, it symbolizes your company's integrity, professionalism, and attitude toward home buyers. Meticulous attention to details creates the reputation you want.

Completion of Warranty Items

While everyone agrees that the best service is not needing service; then the next best choice is fast, courteous, and effective repairs. Therefore, besides studying the nature and number of warranty items, you should track completions. Most builders today recognize that a timely response is a critical component of home owner satisfaction. Because the traditional 30-day completion time is yielding to this pressure, 10 workdays is becoming the common target. Builders need accurate information on completion from the trades. Return of the signed work order provides the best documentation for this work.

Many companies use a community report that lists the trade, the home owner, and a brief description of the items that the trade should correct. Warranty personnel make notes on the printed reports and the administrator uses these updates to prepare the next pending work report, usually on a weekly basis.

You can sort completion data by trade. You should give each trade response-time feedback as shown in Figures 11.2 and 11.3. The company should take action when a few trade contractors consistently fail to meet the required response time. Failure to act sends a dangerous message: "We know these trades are slow, but we are letting them get away with it." On the other hand, when the company replaces a trade whose poor performance has failed to improve after several months, this action also sends a clear message.

Warranty Budget

For many years builders estimated warranty expense at 0.5% of the sales price of the home ($500 for a $100,000 home). In recent years that amount has increased to 0.6 or 0.7%—$600 to $700 for the same home. Some builders set aside as much as 1.5 or even 2% for service. This percentage is more often true on high-end homes. In a few cases, aggressive quality management during construction has resulted in warranty costs as low as 0.3%.

In any event, when warranty costs exceed an acceptable amount, look beneath the surface to discover the causes and work to eliminate them. Excessive warranty costs result from one or more of the factors discussed in the following paragraphs.

Poor Design or Purchasing Decisions. Electing to save $45 per home by using a lower-quality material or method often results in sweeping repairs and replacements during warranty. Worse, the damage to company image can take months or years to overcome.

Inadequate Supervision During Construction. Failure to walk homes and check the work of each trade builds a lot of warranty work into the product.

FIGURE
11.2 Trade Contractor Work Completion Report

Trade Contractor Work Completion Report
Report Criteria

Date Range:	01-01-20__ through 03-31-20__
Trade contractors included:	Top 5 and bottom 5
Minimum number of items:	10
Displayed by:	Total
Expected completion time:	30 days or less

Completion Report Data*
Top 5 Total

Rank	Trade name	Items	Average	Fastest	Slowest	On time
1	Summit Carpentry	35	2D	0H 40M	6D	1500%
2	Frank Concrete	17	6D	2H 05M	9D	500%
3	Pro Insulation	58	7D	3H 25M	12D	429%
4	Topp Flooring	174	11D	6H 05M	19D	272%
5	Superior Finishers	36	17D	3H 05M	32D	174%

Completion Report Data*
Bottom 5 Total

Rank	Trade name	Items	Average	Fastest	Slowest	On time
5	United Millwork	36	53D	9D	94D	57%
4	Miguel's Paint	174	61D	3H 05M	164D	49%
3	Bob's Walls	58	66D	4D	220D	45%
2	Joe's Plumbing	24	85D	12D	167D	35%
1	Al's Cabinets	17	112D	36D	245D	27%

Source: HOMsoft™, Burnsville, Minnesota

Sometimes the cause is staffing at levels that make walking every home as work occurs physically impossible. This problem also can result from having superintendents who lack construction knowledge, time management skills, or organized work habits. To be effective, they must support this effort with written checklists.

Delivery of Incomplete Homes and Failure to Complete the Orientation List. The first perfect home has yet to be built. Anyone can make a list if he or she is motivated. Home owners who are frustrated with slow or no service make such lists, and they are long. Nitpicky items are more difficult to turn down when the house was not "right" to begin with.

FIGURE
11.3 Work Status Report by Trade Contractor

Work Status Report by Trade Contractor

Date range: June 6, 20__ - Sept 6, 20__

Status: Active and completed items

Total number of items: 5 (2 outstanding)

Jiffy Plumbing, Contact: Jeff Jiffy, owner

Binetti · 34452 Valley Drive, Yourtown, ST 99999

Lot: 9 · Block: 1 · Development: Shady Creek

Home Phone: (555) 555-1234 · Work phone: NA

WO no.	Item no.	Age*	Location	Details	Completed date
17254	2	28	Master bathroom	Some sort of apparent leak in master bathroom, either from jetted tub drain or from shower, resulting in stain in ceiling of kitchen, directly below bathroom.	6/28/20__

Clark · 12212 Louisiana Ave, Yourtown, ST 99999

Lot: 4 · Block: 7 · Development: Shady Creek

Home phone: (555) 555-9876 · Work phone: (555) 777-7777

WO no.	Item no.	Age	Location	Details	Completed date
19025	11	15	Kitchen	Fix the plumbing.	7/30/20__

Gundersen · 12345 Harmony Lane, Yourtown, ST 99999

Lot: 6 · Block: 7 · Development: Shady Creek

Home phone: (555) 555-5432 · Work phone: NA

WO no.	Item no.	Age	Location	Details	Completed date
19190	2	4	Master bathroom	Water leak in ceiling of living room, appears to be a leak from water line or drain in master tub. Leak has not been apparent until tub was used recently. ...ceiling and repair water line.	8/7/20__

*Days since work order was issued.

Source: HOMsoft™, Burnsville, Minnesota

Failure to Define Warranty Guidelines for Buyers. The omission leaves buyers to request what they think should be repaired. You can improve this situation through the effective use of a comprehensive home owner guide presented at contract and referred to in planned meetings as described in Chapter 1.

Lack of Training of Warranty Reps. If warranty reps lack training, either in the technical interpretation of the warranty standards or in how to say no to a home buyer, extra expense can result. The more technical understanding warranty reps have

Warranty personnel seldom make decisions about repair items based on the warranty budget. You and your warranty staff should base warranty responses on the physical conditions and circumstances compared with the commitment the company made. Watch the budget not for guidance on decisions about repair items but for direction in improving the product.

and the better their communication skills the more easily they can make correct decisions and explain these decisions convincingly to home owners.

Paying Trades for Warranty Work or Failure to Back Charge. Turnover of trades often results in a building company paying a new trade to correct the former trade's errors. Builders also encounter extra expense when they neglect to levy appropriate back charges. This mistake can occur when recordkeeping is incomplete or when a builder resists back-charging out of a sense of friendship for the company's trades. While this situation is certainly the builder's business choice, it creates an inaccurate impression of warranty costs.

Inadequate Warranty Staff. You cannot assign all warranty items to a trade. Fast response by a well-trained in-house service technician can be a warranty department's best defense against the outrageous lists that grow from slow or no response.

Coding or Arithmetic Errors. Approving bills with mistakes in calculations, for exaggerated amounts, or coding items to warranty when they should be coded to construction inflates the warranty budget.

Data Drives Change

As you analyze items, you may find more opportunities for improvement than you bargained for. If this situation occurs, prioritize these opportunities and select appropriate targets for improvement. Ask yourself the following questions to determine what issue to address first:

- How frequently does the problem occur? In 3% of your homes? Forty-seven percent?
- What is the correction cost in terms of administrative time and repair dollars?
- Is the hassle factor for the home owners high, medium, or low? Consider the number of appointments, dust, noise, and overall inconvenience involved.
- What is the cost to prevent the problem during initial construction?

Small-volume builders benefit from reviewing warranty work, home by home. Treat the information as a case study and search for ideas that could improve future performance. One custom builder looked back on just two items repaired for a home owner but learned the valuable technique of containing the family cat comfortably in a closed room to prevent it from running through the room and tipping over a gallon of wood stain.

For example, if an item (a) occurs in 41% of your homes, (b) is a serious nuisance to home owners, (c) costs $197 to repair, but (d) costs only $61 to prevent up front, it is worth preventing, especially when you factor in the potential damage to your company reputation of not doing so. If an item shows up in 7% of the homes, takes one short visit and $16 to repair, but costs $58 to eliminate up front, you probably would continue to repair it. When you identify a worthy target, summarize your information in a concise memo and suggest a meeting with the people involved to discuss the situation. Once a change is approved, the next step is to implement it fully and check to be sure it is done.

Two questions help you avoid pitfalls. "Who needs to know about this change?" and "What else must be done to carry this policy all the way through the system?" Check all possibilities, including your staff, future purchasers, salespeople, accounting, draftspeople, designers, trade contractors, superintendents, and orientation reps. Information available in model homes and all documents also should be checked, including the contract, home owner guide, sales brochures, blueprints, and specification for trades.

Finally, consider whether any home buyers already under contract will be surprised by a change. Advising buyers of a change and explaining the reasons behind it before they discover it at orientation will prevent conflict. Often, the buyer will happily accept the change; however, if the buyer has an objection, you should resolve the issue prior to the closing.

Service Holes

A service hole is created over time as a builder focuses on selling and building homes but not, necessarily, on warranty service. Service holes are not the product of excessive demands from home owners. They are an accumulation—a backlog—of legitimate warranty items. Service holes can result from a shortage of staff, time, or money. Often they involve a promise to get caught up on warranty work at some point in the future (for instance, as soon as "we get past this year-end crunch")—a point that keeps moving.

Acknowledgment

The first step is to recognize the situation. You need to consider the symptoms of service holes as additional items for your early warning system. Service holes include routine response times of more than two weeks for an inspection appointment and more than four weeks for repairs. Therefore, monitoring the workload of warranty staff also becomes part of a builder's early warning system.

Notice the tone of home owner communications as well. When nearly every communication from home owners carries the sting of insult, when lists received by warranty are several pages long and include many complaints about orientation items that were never corrected, when large numbers of home owners call in third parties (attorneys, Better Business Bureau, private home inspectors, their real estate brokers, and so on) for support—*that* is a service hole.

You cannot tack the assignment of getting the company out of a service hole onto normal warranty staff duties. Builders cannot climb out of service holes with the same systems that got them into them in the first place. This special situation requires a special plan and commitment. Someone needs to work at the job full time with no distractions. That someone may need the support of an administrator and one or more technicians—a service task force.

Begin at the Source

Make necessary changes to stop the oversupply of warranty work. A common culprit is delivering incomplete homes—the warranty staff inherits already angry home owners. As warranty staff devotes more time to handling hostility and addressing incomplete orientation items, less time is available for normal warranty work. This situation creates a self-perpetuating cycle. By gaining control of a warranty overload at the source, you allow the task force a realistic chance to catch up and cure the situation.

Have a Plan

Meet with home owners one by one. Listen, inspect, and make an honest estimate of the time for repairs. Inspections can be done faster than repairs. Organizing a schedule of certain days of the week for certain activities, or splitting work days into an office half and a field half may work.

Resist the temptation to do extras in an effort to "make it up" to customers. An organization that was having trouble providing the basics is not in a position to do extras. Making home owner Jones wait for something he paid for, while home owner Smith gets something she didn't pay for, only makes things worse.

Calls and E-Mails. Service holes get bigger when staff is tied up on the phone or with e-mails to the extent that they can't get anything else done. Yet customers cannot be ignored; they deserve to have their questions answered, their complaints aired. Designate one primary person (have a backup for breaks and lunch) to manage such home owner contacts while the rest of the team is out in the field meeting home owners, doing repairs, or contacting trades.

One company facing a service hole established a team to address the issue. They referred to the "back log" and "front log" to distinguish the task force assignment from normal warranty operations. They challenged the back-log team to complete its assignment and disband in 90 days. A dinner celebration held when the assignment was completed early—at 81 days—was a point of pride for the entire department and the trades who helped.

The phone person can set up inspection appointments and, between calls, look up information needed by the field team. Often the builder struggling with service is the same one whose records are in disarray. Which trade contractor did which house? What is the paint number on that home? Where can we get screens to fit these nook windows? Searching for answers to such questions takes time. With phone and e-mail support and the needed information, progress will occur faster.

Trade Support. The emphasis on warranty work may startle some trades. Explain the situation in a series of phone calls or with a memo expressing your commitment to resolve it. Point out that meeting this challenge is in the best interests of everyone involved.

Expect a Surge. When home owners begin to see that your company really is taking care of warranty items, expect an increase in their input. The home buyers who had given up hope will step forward to report their items (in some cases for the second time). Stick with it. Inspect and provide service as needed in accord with normal standards. Do what's right—one home buyer at a time, one detail at a time. Monitor progress, including a review of trade contractor performance, at least once a week.

As time passes and the pressure begins to lighten, you will see progress. You can then turn the service energy toward making sure a backlog situation doesn't happen again. Establishing appropriate customer expectations, quality control, complete deliveries, and efficient warranty practices are all easier on budgets and nervous systems than crawling out of service holes.

Optional Reports

With the basics under control, you may want to expand your reporting activities. Special reports might be produced temporarily, intermittently, or on a one-time basis to gather details about a unique situation.

Denied Items

Tracking the work you agree to may seem like plenty to handle. However, studying warranty requests you deny also can be useful. For instance, if 43 out of 60 home owners ask for the same repair only to have it denied, that might be a good repair to provide. If providing the repair is impractical, look for a way to set buyer expectations more clearly in the beginning.

Warranty staff often can identify a salesperson who may need more training in aligning customer expectations. By taking note of which communities generate home owners with inappropriate expectations and the subjects of those inaccurate expectations, warranty staff can provide insight that helps salespeople refine their presentations to buyers.

Hot Files

The potential lawsuit, the infamous home owner with a vile temper, a material failure affecting several homes, or a serious structural repair are examples of cases that might be assigned to hot-file status. Review the status of each of these situations with upper management (or the company owner), at least once a week.

Warranty Administration

One of the hidden delays in warranty work is administrative time. Contact home owners within one business day to set up an inspection appointment, or issue work orders within that same time frame if an inspection is unnecessary. Consider tracking—perhaps temporarily or intermittently—the time span for three administrative aspects of warranty work:

- From receipt of a service request to setting up an inspection appointment
- From receipt of a service request to issuing work orders if an inspection is unnecessary
- From inspection appointment to issuing work order(s)

This tracking offers another opportunity to establish one or more early-warning-system criteria. The recommended response time for each of the activities listed above is between four hours and one business day. If any of them takes longer, find out why.

Home Owner E-Mails and Calls

Some builders benefit from tracking the number of e-mails and phone calls, broken down by type as follows. If any particular category seems to generate an excessive number of contacts, investigate that category:

- Emergency reports
- Feedback about a trade or in-house service tech
- Complaint about the quality of work
- Complaint about a missed appointment
- Questions about the home—features, materials, colors, and so on
- Incomplete orientation items

Winter Work

In some climates, builders may close homes during the winter with exterior items incomplete. The promise to return in the spring to complete these items can cause home owner hostility unless you track those items well. Make this tracking

If the number of phone calls seems inexplicably high and especially if those calls are from home owners who want to report lists of items by phone, look for opportunities to improve buyer education about how best to report items. Time spent on the phone recording lists of items is time you are not spending getting repairs accomplished.

FIGURE
11.4   Postponed Winter Work

[Logo]

Postponed Winter Work

Home owner and address	Closing date	AC	Final grade	LS	Concrete	Paint/ stain	Other
Smith 3365 West Port Ave	2/28/20–	X	X				
Raymond 3387 West Port Ave	3/2/20–		X				
Zales 615 West Port Lane	3/14/20–	X	X		X		
Isaacs 622 West Port Lane	3/15/20–		X		X		

convenient by gathering winter work information from orientation forms and compiling a postponed winter work list (Figure 11.4). At the appropriate time, use e-mail, fax, phone calls, letters, or postcards to let the affected home owners know they have not been forgotten.

Format and Timing

Reports should be readable, consistent, and timely. Begin with the report title and date at the top of each report. Decide on an appropriate format for the data. Whether you are reporting statistical data in tables or editorial comments from home owner surveys in a text format, present the information in a logical order with clear headings and subheadings to make it understandable. Beware of reports that duplicate effort without producing any new information; too many reports can be as problematic as too few.

By establishing set time frames for producing reports, they become a normal part of warranty routine as described in Chapter 12. Most reports are compiled on a weekly (work order completions) or monthly (recurring items) basis. Some are more appropriately done quarterly (expenses). Channeling accurate information to architectural, land development, purchasing, construction, sales, and selections personnel can take your company to new levels of excellence and increase profitability. Warranty holds up a mirror to the earlier stages of company performance. Wise companies take advantage of the available feedback and reap the benefits.

Warranty Office Routine

An effective warranty office handles dozens if not hundreds of details, resolves issues promptly, quickly provides appropriate repairs, and documents every step along the way. Home owners are impressed, staff members enjoy their work, and the company sells more homes as a result of well-managed service. The question is "How does the company achieve that idyllic condition?" Efficient work habits have always been valuable. In today's fast-moving, ever-changing, information-abundant environment, such skills are vital to survival. However, many people today find themselves moving so fast that they miss opportunities, overlook relevant details, or make foolish mistakes. Do any of the following examples sound familiar?

* A service administrator overlooks a note from a home owner that he's home on Thursdays and by doing so makes completing the repair work more difficult for all involved. The home owner and the trades affected conclude that the service staff is either uncaring or incompetent.
* A warranty manager fails to note an appointment with an already difficult home owner and schedules a meeting with a trade for the same day and time. She realizes the error when the trade mentions the same home owner during the meeting. The warranty manager dreads calling to apologize and makes matters worse by putting off making the call to reschedule.
* Accounting issues a check to pay a trade for courtesy work in a home under warranty. The bill is for 28 hours of work at $18 per hour, performed by one technician, all on the same date. In approving the invoice, the warranty rep failed to read and question the details. Had he called the trade, they would have uncovered the error: the decimal point was missing. The bill should have been for 2.8 hours—a difference of $453.60.

By taking a bit more time to do a task right the first time, you will avoid most careless errors like these. Not only will you work more effectively, but you also will enjoy your work (and life) more if you avoid working at the speed of mistakes. Create a sense of order by establishing a supportive work environment, efficient work habits, and a solid routine for normal warranty activities.

Create a Positive Work Environment

Confronted with multiple tasks, first consider the physical setting for the work. Good habits will generate better results in a functional environment. Taking time to organize and maintain both office and vehicle will improve service.

Office

At a minimum, warranty work requires quiet, privacy, high-quality equipment (phone, fax, computer, printer, and copier), and space to store materials and files. If your warranty office does not meet these criteria, lobby calmly but persistently for improvement in both space and furnishings.

The quality of the space is as important as the amount. Noise, lack of privacy, interruptions unrelated to warranty work, or a dark, cheerless environment diminish productivity. One warranty staff experienced an attempt at decorating: walls included oil paintings of ships in a stormy sea—just the right touch for a warranty staff drowning in complaints.

"Excuse me. . . ." "Pardon me. . . ." "May I get through please? I just need something from that file. . . ." Hundreds of times a day dozens of warranty staff people squeeze, twist, climb, bend, and contort themselves into cubby holes sadly referred to as their offices. Consider your physical setting: Does one person need to stand up for a second person to enter the room?

If one person needs to put a finger in one ear to hear what home owners are saying over the phone, changes are needed. Consider the impression made on the home owner who hears bits and pieces of background conversations while attempting to describe a roof leak. Décor might include a large bulletin board for posting the latest customer satisfaction ratings, letters from home owners, and relevant company announcements.

Organize Your Desk

For the record, a clean desk is *not* a sign of a sick mind. Keep your desk tidy. While having everything in its place saves time and stress, clutter is distracting and encourages losing items. Have a system in your work area. A standard place for items you need to read, copy, approve, file, or otherwise address helps you to avoid losing things. You can also see at a glance how much of which type of work is waiting for your attention. If you had all the time you spend looking for something, you would get a lot more done in a calmer frame of mind.

Does your work area support efficiency? Is it organized and supplied to help you work with a minimum of hassle? If you see room for improvement, consider these suggestions:

* Keep desk top tools within easy reach and maintain a clear work space in the center.
* Maximize your inventory of supplies and frequently used forms. Store the most-used items as near as possible to where you use them.

* Create visual "systems" on your desk—phone calls to return; calls you made that await responses; bills to approve, copy, file; and so on.
* Batch such items and keep them in the same place all the time. Keep like things together, unlike things apart. Avoid "mystery stacks."
* Use color coding to simplify filing.
* When creating a new file, ask, "Where would I look for this?" instead of "Where should I put this?"
* Do any of your tasks lend themselves to tickler files with titles such as days of the month, months of the year, staff meeting, and so on?
* Establish legal dates for disposing of records, and store duplicate backups of electronic records off site.
* A bulletin board is a good place to keep frequently used items such as the company phone list or pending items, such as purchase orders, when you are waiting for deliveries. File signed purchase orders in chronological order (latest on top) in a 3- or 5-inch, three-ring binder in addition to the unsigned electronic copies stored online.
* Check sound, temperature, and lighting—do any of these environmental factors interfere with performance? Is your desk chair comfortable and at the proper height for inputting items into your computer and reading the screen?
* Archive unused materials, such as files for home owners more than one year out of warranty, and budget printouts that are more than a year old. Organize them with like things together and label them clearly in case you ever need them again.
* Evaluate your technical skills with the phone system, pager, handheld computer, printer/scanner, laptop, and postage meter. If you lack some skills, commit to mastering one skill at a time.

Vehicle

Warranty personnel typically spend time in the field each week. Inspections, home owner follow-up, meeting with trades, or attending community team meetings are a few of your out-of-office duties. Ensure maximum productivity by stocking your vehicle with helpful items.

* If you carry files in your vehicle, use a plastic case or file box to keep them neat and prevent losing items.
* Install a notepad on the dash board for taking notes while stopped in a traffic jam.
* Use a small, voice-activated recorder to tape reminders while still driving safely.
* List items you typically need to meet with home owners or one of the trades (business cards, work orders, purchase orders, tape measure, level, and so on) and stock your vehicle with them. If many different items are involved, create an inventory checklist and use it regularly to restock your supplies.

Develop Productive Work Habits

An efficient environment permits, but does not guarantee, effective work. Productive work habits and self-discipline are essential to managing warranty work. Start with the following habits:

* Determine which time of day is your golden time—the hours when you are the most alert, well-focused, and most productive. Then schedule your most challenging work for that time of day. Leave more routine tasks to the remaining hours.
* Plan your next day at the end of the current one or at the beginning of the next. Determine which works best for you. One hour of planning typically saves three to four hours in execution. Keep an eye on your horizon so big projects do not sneak up on you. Anticipate future events and plan for them.
* On busy days, remind yourself to stay calm and focused. Your goal is to move quickly with a calm mind. If you find yourself dropping things or being unable to do common tasks, take a breath and calm down. Double-check that you are indeed putting your time on your priorities. Concentrate on each task, however minor. The goal is to deal correctly with each issue and avoid having to perform the same chore twice, having to apologize for errors, or being unprepared for an event you care about. Recognize when you are operating in frantic mode. Then pause to calm down and refocus. Because anxiety produces short, shallow breathing, deliberately take deep breaths periodically to help keep you calm.
* Always have something to write on and write with. The fast pace and large number of details to track promotes forgetfulness.
* Take time to read company memos, letters, and reports carefully. Keep removable, stick-on notes handy in case you want to record some thoughts without writing on the original copy.

Organize Your Planner

An appointment book or electronic planner is one of your most valuable time management tools. Customize a purchased system (paper, wall, computer, or handheld computer) for your particular needs. Discover the formats and methods that work best for you. To use your planner well, determine what it should contain to support your daily work. Once you have tailored your planner to suit your needs, use it effectively, as follows:

* Use just one. Keeping two or more calendars updated and accurate is nearly impossible and takes up too much time. If someone needs a copy of your schedule, make a copy of your original one at appropriate frequencies—daily if needed.
* Keep it with you at all times while working.
* Make a space for your planner on your desk—one where you can easily see and reach it to make additions or retrieve information.

* Reduce floating notes by recording tasks or ideas in the appropriate place immediately.
* Note the person's name and phone number on the date you have an appointment.

Communication Guidelines

Phone calls and their many modern variations—faxes, e-mails, memos, and correspondence—abound in today's business world. Manage this flow of communication effectively for responsiveness and productivity. Remember that being immediately accessible at all times can seem like sound customer service on the surface, but it deteriorates when you are so overloaded with interruptions that you cannot accomplish anything for anyone. One study concluded that if you interrupt people every 20 minutes you reduce their productivity to zero. Your challenge is to find a balance between being accessible and producing results. The following guidelines will help you manage communication:

* Set expectations with all of your customers (home owners, trades, associates, and colleagues) about contact with you. Give them the best way to reach you (day, time, and method) and likewise learn the best way for you to reach them.
* Establish an expectation of an acknowledgment within one business day and a personal standard to respond to or at least acknowledge a contact within four hours.
* When you are unable to resolve an issue in one contact, estimate the time needed and commit to a follow-up date. Record this date and keep the appointment even if you are still working on a final answer: "I'm just calling to tell you I have nothing to tell you." Then schedule the next update appointment.
* Excepting real emergencies, batch communications whenever possible. For example, return calls and respond to e-mails once in the morning and once in the afternoon.
* Set up "talk to" agendas for the people you communicate with regularly— whether by phone, e-mail, or in person. Accumulate a list of subjects and cover all of them in one call to each person. They will appreciate not being interrupted repeatedly and save both of you time.
* Respond to short memos with notes in the margin and then make a copy for yourself if you need one.

Phone Calls. Efficient phone-use habits can increase your productivity and the levels of home owner satisfaction dramatically. Consider which of these suggestions listed below might be useful:

* Make reasonable use of voice mail or an answering service. People do not mind leaving a message if they know you call back in a reasonable time (target one to four hours).

- Check for messages regularly—avoid hiding behind your voice mail. Likewise, check your e-mail twice or more each day.
- Create a separate phone message log for each call (Figure 4.11).
- Prioritize your calls.
- Plan points to make in each call. A planned call takes an average of 7 minutes; an unplanned call takes 11 to 12 minutes.
- Document the date, the time, and the points you made. Use standard abbreviations to make this procedure move faster.
- When you place a call, be ready to leave a message—nearly 90% of the time, the person you call will be unavailable; half of the time, the two of you do not need to talk directly anyway.
- In all phone messages, include your company name; department, if applicable; first and *last* names; phone *and* extension number.
- If you need a callback, say so and provide a couple of times you will be available.
- Have several techniques for ending a conversation once the business of the call has ended, such as, "It's been good talking to you," or, "Thanks for the update." Or summarize the decisions made in the call.
- Document each call immediately—mark your calendar and so on.
- When you are unavailable for an extended period of time, arrange to have someone else cover for you—at least acknowledge your callers and explain when you will return.
- Avoid making commitments for others without their consent. "I'll have him call you this afternoon" sets up a customer for disappointment. "I'll give him the message" is more accurate unless you know for certain the party you are referring to is available and willing to follow up.

E-Mails. As with phone calls, e-mail must be managed consciously to be part of an effective communications strategy. The following simple steps can make e-mail usage more productive:

- Keep in mind that you may be required to present all e-mails in a courtroom through "electronic discovery." Therefore, keep personal opinions in your head and out of your computer. Don't write anything that you wouldn't want to present to a judge.
- Confirm that complete and accurate contact information automatically appears at the bottom of every e-mail: your full name, title, company, phone and extension numbers, fax number, and mailing address.
- Avoid using your "in box" as a staging area. Establish "save" or "print" folders for email you want to follow up on.
- Some e-mails may need to be printed so you can add notes, carry details with you, or file.
- In e-mail, avoid using ALL CAPS—the typographical equivalent of shouting.
- Compose complex or emotionally charged e-mails off line and paste them for sending once you have edited the content.

Faxes. While faxes may not be as ubiquitous as are phone calls and emails, nonetheless you must manage them effectively.

* Include your contact information clearly and accurately on a fax cover sheet along with the total number of pages you are faxing.
* Avoid assuming that because you faxed an item it was received. Leave a message that the fax is on the way.
* When sending a long fax, call first to alert the receiver and make sure the recipient's machine is prepared to receive it.

Letters

As mentioned earlier, assemble standard form letters and examples of favorite letters addressing unique situations. If you find yourself struggling with where to start, focus on the question "What's the point?" Then jot down your thoughts, identify the main points, add supporting details, and arrange everything in a logical sequence. Frame your material with a brief introduction and a brief conclusion. "This letter is to confirm the conversation we had on Tuesday. . . ." Bring your letter to a conclusion in a businesslike tone. This conclusion can be as simple as inviting the customer to call you with further questions. Remember that customers are less likely to read long, wordy letters than brief, concise ones.

Plan for Interruptions

Include a margin in your daily schedule for the inevitable interruptions that are part of warranty work. Avoid scheduling yourself so tightly that any emergency destroys your entire day. Then manage interruptions effectively, eliminate those that can be eliminated, and quickly handle those that cannot. To reduce interruptions in your office, consider the following tips:

* Have a visual barrier to prevent you from making eye contact with people who pass by. For instance, having two work surfaces might allow you to turn your back to the door.
* Have few extra chairs in your office—at least near your desk.
* Leave—go to an unoccupied conference room, an empty office, or the library for quiet time. If they can't find you, they can't interrupt you. But leave word of your whereabouts with a colleague in case of an emergency.
* When you are interrupted, taking a moment to make a quick note to show where you stopped can help you return quickly to the original task.
* If you need to talk to someone who is typically long-winded, go to that person's office so you can more readily control when the conversation ends.
* Be ready for people who ask: "Do you have a minute?" "Actually, I have about two minutes." Or, "I have some time open at 4:30. Will that work for you?"
* Do you show your staff and associates that you respect *their* time? What interruptions do *you* cause? Evaluate your answers and improve your behavior where you can.

Manage Projects

Warranty often has special issues to study and resolve. Such projects might include learning new software, remodeling your office or moving to new space, closing out a community, and preparing the model homes for delivery to purchasers. Your challenge is to manage the project and keep up with all routine tasks. To accomplish both goals, use the following tips:

- Define the project clearly.
- Be certain you understand the desired results and time frame.
- Brainstorm the probable steps involved.
- Put the steps in sequence.
- Expect to think of more ideas after you have had time to reflect.
- Identify storage space and a tracking method—a project checklist, a chart on the wall, or some other method.
- Allow time for research.
- Schedule blocks of focused time. Completing one step creates momentum.
- Identify additional resources you will need: information, people, and equipment.
- Schedule backward from the deadline. Include a contingency time buffer for unexpected events.
- Can you delegate parts of the project work or some of your routine tasks while you're involved in this project? If so, do so.
- Along the way, take notes of steps you did not anticipate and update your project checklist, wall chart, or whatever scheduling system you are using. File this information for future reference—the next time something similar comes along, you will already have the beginnings of a plan.

Learn to Say No

Others will use up your entire workday if you let them. Each time you choose to do a task, you are choosing not to do others. Taking on too much can result in doing all tasks less effectively and disappointing everyone. If you decide the task is not for you, say so:

- "I appreciate your asking. I'm committed elsewhere. I'm sure you'll do well with that."
- "Sounds like a great project. My schedule prevents my doing justice to it. Have you asked Harry to work with you?"
- "Which is more important, A or B? I can complete one of them by the end of the week."
- "If you can help me with this project, I can help you with that one."
- "I could get involved with that in March. At the moment my calendar is full."
- "I'm sorry, but I'm unable to do that task."

Delegate Tasks

To delegate effectively requires first that you completely understand the task so you can describe it accurately. Maintain a delegation log—the who, what, and when of tasks you have delegated—that shows when updates are due. When you delegate a task, be prepared to explain it. Some examples follow:

* Describe the results you want and by when. Establish measurable success/completion criteria. Suggest sources and possible approaches, but not methods. Delegate the what, not the how.
* If interim progress reviews are appropriate, schedule them at the beginning.
* Make deadlines realistic. Clarify that the person to whom you are delegating must bring any delay to your attention immediately. Avoid delegating a task at the last minute when no one could do a good job.
* Point out why you selected this person. (Select someone who has the necessary skills and wants the responsibility.)
* Establish boundaries of authority. If the task you are delegating is to tighten warranty response times by the trades, you might say "If you get to the point that you believe holding someone's check is the only action that will get results, discuss the situation with me first."
* Provide feedback and appreciation—express your appreciation or provide a tangible reward upon completion of the assignment.

In the Field

When you're out of the office, whether you are meeting with company personnel or performing home owner inspections, let the receptionist know when you will be available or when you will check in for messages. Update your voice mail response message. These practices earn a lot of goodwill with the person at the front desk and can prevent you from receiving multiple messages from the same caller. In addition, you may want to follow these tips:

* Arrange your stops in a logical order.
* Use waiting time to plan or carry a folder of professional articles you want to read.
* Listen to industry tapes in your vehicle to make good use of travel time.
* If you need to make a phone call, pull over and park in a safe location. Talking on the phone uses the same part of your brain as you use for driving. Don't put yourself in danger of having to choose which activity to attend to.
* Store business cards in the same place—yours to give out and those you receive. Get in the habit of carrying your business cards in the left pocket of your suit jacket and putting business cards you receive in the right. This arrangement allows you to hand your card with your left hand while your right hand is ready to shake hands or receive the other person's card.

* Store receipts you will need to fill out an expense report in the same location to avoid having to search for them.

Improve Warranty Routines

In addition to managing the constant flow of communication that makes up warranty work, establishing regular times and habits for typical warranty office (or desk) tasks increases productivity and reduces stress. Increase your efficiency by creating checklists and inventories for tasks that involve multiple steps. Besides ensuring that you cover everything, these tools give you a place to note where you stopped if you are interrupted.

An overview of routine warranty activities such as the one shown in Figure 12.1 allows you to keep upcoming tasks in mind while addressing daily details. Some of the activities included are discussed in detail in the chapters referenced in parentheses; others are highlighted here. Depending on company volume and staff levels, routine tasks can vary significantly from those shown in the sample. But it demonstrates how capturing repetitive activities in this manner is a good starting point for developing an efficient routine for warranty activities.

Administrative Tasks

The administrative heart of any warranty office is the written work order. While they are easy enough to issue, especially with an effective computer system, managing the resulting flow of information and retaining adequate documentation can be challenging. To meet this challenge, ensure that the warranty process provides you with an adequate framework.

New Home Owner Files. The house closing is the signal to create the warranty file. Establish a file for each home and speak with an attorney about a document retention policy. At a minimum, documents should be kept for the period of the statute of repose or the statute of limitations in your state. Standardize the information you include on the warranty file labels: Home owners' names and addresses, closing dates, job numbers, and models. Use color-coded file labels to make identifying the community where the home is located easier and misfiling less likely.

A warranty file should contain the following:

* selection sheets and change orders
* the orientation list
* home owner requests for service
* inspection reports
* completed work orders
* copies of correspondence
* records of phone calls
* photos, diagrams, and other related material

File items in chronological order. If each new item is placed on top in the file, anyone can remove a file folder and read from back to front to review the home's service history. When the warranty expires and year-end work is complete, you may want to remove the file and store it separately from the active warranty files, but not out of reach.

Reports. Reports and their uses are discussed in detail in Chapter 11. As valuable as they are, reports need to be produced and circulated in a timely manner for them to be useful. To fit them into your warranty routine, consider the following questions:

* What reports do you need?
* At what interval? Most reports are compiled on a weekly (work order completions) or monthly (recurring items) basis, although quarterly may be more appropriate reporting for some (budget, for example).
* When is the report due?
* To meet the schedule, what cutoff date do you need for the data included in the report?
* Who will compile the report?
* Who (besides you) will receive the report?
* When will you discuss the report and with whom?

Budget and Bills. If warranty authorizes payment to a trade contractor or supplier for a home owner, the bill should be reviewed, approved, and coded by the person familiar with the work ordered. Like other routine tasks, this one is more likely to be done in a timely manner if you set aside a regular time to process invoices.

* Be familiar with the individual trades' scopes of work so you know what you should pay for.
* Are time and material costs documented? Was there an estimated cost for the assigned work?
* Learn what information accounting needs and provide those details on every approved invoice. This practice avoids delays and extra work. It also ensures that your trades get paid on time. Accounting usually needs at least a defect code, a community name, and a lot or job number.
* Track warranty expenses by home rather than collect them into one large warranty category. The costs of completing the orientation list typically is applied to construction's budget. Costs incurred for items discovered after closing belong to warranty.
* An "out of warranty" designation is appropriate to categorize expenses for analysis. If your system does not provide for this line item, ask for it to be added.

FIGURE

12.1 Warranty Office Routine (for more details, refer to chapters or figures noted)

Task	Daily	Weekly
Home owner contacts		
Calls, e-mails, and faxes	• Emergency: Respond immediately. • Nonemergency: Respond within 4 hours. • Set up/confirm appointments. • Follow up on issues or questions.	• Follow up with home owners and trades regarding work orders: incomplete and completed (Chapters 7 and 8).
Inspections	• Conduct inspections (Chapters 4 and 5).	
Letters and home owner literature	• Compose individual letters to home owners within one business day.	• Send welcome to warranty letters. • Send year-end letters (Chapter 4).
Administrative tasks		
Work orders	• Issue work orders within one business day of inspection (Chapter 6).	• Pull or log out completed work orders.
Files		• Set up files for closed homes. • File home owner documents.
Reports		• Update pending work order report for staff meeting. • Review hot files.
Budget		• Approve invoices.
Supplies and equipment		
Personnel and associates		
Warranty staff communication and training	• Answer questions/discuss issues as needed.	• Conduct warranty meeting (Figure 12.2). • Invite other department reps to join meeting. • Conduct training segment at staff meeting (Chapters 2–10 and 14).

FIGURE
12.1 *Continued*

Monthly	Quarterly	Annual	As needed
Home owner contacts			
			• Check with the home owners on roof repair list after significant rain. • Respond to severe weather damage reports (Chapter 7).
			• Assemble community binders (Figure 5.2).
	• Send seasonal maintenance hints in letter or newsletter (Chapter 13).	• Update home owner guide (Chapter 1).	
Administrative tasks			
	• Pull/archive expired warranty files.		
• Compile/print warranty item analysis reports.		• Develop budget for next year.	
	• Review budget versus actual costs.		
• Inventory office supplies and forms. Replenish as needed.		• Review warranty office space and work environment.	
Personnel and associates			
• Arrange for manufacturer training workshop (Chapter 14).	• Review staffing needs versus workload (Chapter 14). • Visit competitor/report findings at staff meeting.	• Plan training (Chapters 2–10 and 14). • Conduct performance appraisals twice each year (Chapter 14). • Read all customer documents.	• Schedule new warranty employee orientation (Chapter 14).

Continued

FIGURE 12.1 *Continued*

Task	Daily	Weekly
Personnel and associates		
Other departments		• Attend a community team meeting.
Trades		• Follow up with trades on outstanding work orders (Chapters 7 and 8).

- Check all arithmetic and confirm that the amount charged is reasonable for the work assigned.
- If an invoice contains errors or omissions, contact the trade contractor or supplier immediately to clear up the problem while the information is fresh in everyone's memory.
- Ensure that your accounting system tracks warranty expenses to know the company has met any insurance deductibles.

Although warranty decisions generally should not be made based on how much money has been spent so far, regular review of actual expenses compared to budgeted expenses can call attention to trends that need attention. Include this study on your master schedule as a routine part of the warranty department's work.

Personnel and Associates

Besides maintaining healthy relationships with home owners, working relationships with fellow staff members and associates are essential to effective warranty service. This work requires daily communication within a context of mutual respect and appreciation. Planned meetings, training, and proactive contact with others can contribute significantly to satisfying home owners long term.

Warranty Staff Meeting. The warranty staff should meet regularly, usually weekly. For small-volume builders this may mean that the individual who oversees warranty (along with other responsibilities) meets with the company owner. If you are a "one man" or "one woman" show, set aside this time to review warranty work and take follow-up actions. To ensure that your staff meeting is efficient and worthwhile, standardize the agenda (Figure 12.2).

FIGURE 12.1 Continued

Monthly	Quarterly	Annual	As needed
Personnel and associates			
• Attend a staff meeting: sales, selections, construction. • Inspect model homes.			
• Prepare trade performance report. • Send evaluations of trade contractors to purchasing (Chapter 10).		• Organize social activity with the trades (Chapter 10).	• Conduct new trade orientations (Chapter 10).

As with repair appointments, set standards with your staff for how you will conduct these warranty staff meetings. Some suggestions follow:

- Meet on the same day and at the same time weekly.
- Be prepared. If you or another staff person will distribute materials, have sufficient copies. Review notes from the previous meeting to check for topics for which you or another staff person need to provide an update.
- Remind the receptionist about the staff meeting.
- Start on time and end on time.
- Turn phones, pagers, and radios off unless someone is waiting for emergency information.
- Proceed through the standard agenda in an orderly fashion.
- Allow everyone an opportunity to contribute.
- Avoid sharing war stories unless they are relevant to the topic under discussion.
- When discussing problems, offer solutions. Avoid negativism and whining.
- Record conclusions, decisions, and assignments along with who will be responsible for implementing them and when an update is due.

Status of the Department. One of the most important tasks a warranty service manager addresses is keeping his or her boss informed. A monthly state-of-the-department memo, limited to one page, can provide an efficient update without necessitating a meeting. Should something need discussion, you can simply set up a time to get together. Updating the previous memo helps the warranty manager maintain focus and prevents items from stagnating.

FIGURE

12.2 Warranty Staff Meeting Agenda

Date ___8/10/20–___

Attendees ___Harry Beasley, Alicia Alexander, Ron Smith, Anna Blackstone, Brian Taylor___

Company News

Communities

- Community meetings
- Sales staff
- Model homes
- Amenities
- HOA/common areas
- Grand opening
- Closeout
- Construction staff
- Product changes
- Selections/change orders
- Orientations

Warranty

- Emergencies
- Inspections
  - 60-day
  - 11-month
  - Miscellaneous
  - Out of warranty
- Pending work orders
  - Current
  - Expired (over 10 days)
- Trades' performance ③
  - New trade orientation
  - Evaluations
  - Back charges

Competitor Visit ①

Accounting

- Approved invoices
- Expense statements
- Gas credit cards
- Van maintenance
- Budget review

Technology

- Phones, pagers, radios
- Computer system

Staff

- On-call schedule ②
- Vacation schedule
- Training
- Uniform shirts and blouses

Upcoming Training Schedule

- Review of home owner guide ④
- Manufacturer rep
- Trade workshop
- Other

Challenges and Suggestions for Improvement

Notes

① Alicia will visit Brownstone Village and report

② On-call to be extended to 2 weeks to allow flexibility for vacations

③ Purchasing will replace painter soon due to manpower issues; update from Harry at next meeting

④ Home owner guide review of concrete at next meeting. Contractor will join in to answer questions.

Relate to Other Departments

Staff members in other departments may hear home owner complaints about warranty service. Such complaints might center on response time, quality of repairs, denials, or offenses committed by service personnel—in-house or trade. Use your public relations skills to counter this negativity. A sales or construction person who has confidence in your department can field complaints more effectively. Provide regular updates about your work, including facts and statistics. Show how your efforts improve product quality and customer service. Examples of ways to share information follow:

● Present 5 to 10 minute updates or "infomercials" at sales meetings.
  – Review effective use of the home owner guide.
  – Role play specific service questions or issues.
  – Offer an overview of trade contractors' performance and what you are doing to improve it.
  – Present guidelines on how to handle off-hours emergencies.
● As described in Chapter 1, attend weekly community meetings between sales and construction staff to stay informed about new buyers and their homes and to share feedback from warranty work.
● Improve the quality of service materials. Make certain all sales sites have accurate and current service information.
● Invite staff from other departments to observe some warranty inspections.

The amount of planning required for effective warranty service may seem excessive. However, many warranty service managers (and their companies) have paid a high price for haphazard warranty processes. Setting up reliable systems and developing effective habits from the beginning minimizes conflict, prevents turnover of personnel, reduces costs, and maximizes customer satisfaction.

The Service Horizon: Beyond Traditional Warranty

CHAPTER 13

Most builders believe they have healthy relationships with their home owners. Many are surprised to learn, as mentioned in Chapter 4, that by the end of the first year in their homes the average of home owners' satisfaction has typically fallen 8 to 11% from what it was shortly after move-in. The good news is that this common decline in home owner satisfaction is neither inevitable nor difficult to prevent.

Begin by avoiding any hint of *buyer abandonment* or the sudden end to the excitement and attention to which home buyers have become accustomed. Meetings, phone calls, e-mails, site visits, questions, payments, decisions, and more meetings. This flurry of activity during the building process abruptly stops. Warranty programs that lack proactive procedures convince home owners to think after closing that their home building company has turned its back on them.

By planning ways to stay in touch with home owners, you prevent them from thinking you have abandoned them and promote their satisfaction. The desire for repeat and referral sales alone provides sufficient reason to address this issue, but other factors add to the urgency.

- Failure to provide prompt and effective warranty service may negate the legal benefits of notice and right-to-repair laws.
- Insurance requirements and rising premiums add yet another motivation because builders want to keep their money rather than paying it to insurance companies. With better service and fewer arbitrations/litigations, their premiums are less likely to increase.
- Future home buyers are increasingly aware of home owner satisfaction rankings and are more likely than ever to consider them in selecting a builder.
- When services fail to please home owners, daily interactions with company personnel generate damaging stress and can lead to expensive, time-consuming staff turnover.
- The effectiveness of formal advertising is diminishing although the cost remains high. Consumers rely more and more on referrals from people they know. (Experience shows that this trend is especially true among the expanding market of buyers from other cultures—Asian and Hispanic home buyers, for instance).

Wise builders recognize the potential positive impact of impressive after-move-in services on all of these factors. They take action to ensure that their companies exceed customer expectations rather than disappoint them. Home owners and builders alike can reap the benefits of this effort. To do so, begin with a clear plan for enhanced after-move-in services. **Caution: Always consult with a local attorney concerning the legal consequences of adding or changing warranty provisions.** Consider adding to traditional warranty with ideas from one or more of the categories listed below:

* transitional attention related to move-in activities
* enhanced warranty coverage
* maintenance service options
* informational support
* social activities

Once you have selected and implemented after-move-in extras, confirm that you are providing services that are meaningful to your home owners and that you are doing so effectively. A comprehensive feedback system can help you identify opportunities and trends. Finally, put the best results of your service efforts to work with a planned referral program.

After-Move-In Service Objectives

Adding thoughtful service elements that go beyond traditional warranty practices requires creativity and imagination, empathy and sensitivity. A second caution: Before expanding your services, make certain you are doing the basics correctly: Align expectations, communicate during the building process, deliver the home complete and clean, and respond effectively to warranty requests. Home owners take a dim view of companies that offer frosting when they do not even deliver a cake. Confirm that the fundamentals are in place as described in earlier chapters before you add extras. With the basics under control, think through potential new ideas carefully:

* Define your company's objective(s). For instance, do you want to reinforce satisfaction and stimulate referrals, differentiate your company, develop a profit stream, or some combination of all these ideas?
* Anticipate practical issues related to any service enhancement idea, such as staffing needs, hard costs, administrative support, satisfaction guarantees and liabilities, printed materials, and storage space—to name a few. You have seen the results of governmental mandates that lack resources to implement them at all levels; avoid this error and match service capacity to your commitments.
* If you expect home owners to pay for service extras they select, decide how you will collect and process payments. If applicable, examine the impact these services will have on trade contractor workload.

* Establish criteria for success. If customer satisfaction is the target, establish a benchmark and incremental goals. If increasing referrals is your objective, decide how you will measure that increase. If an income stream is the goal, set an amount to define success.
* Check how your services or the services you are considering compare to those offered by competitors. If other companies can readily copy your ideas, their marketplace value will have a short life span.

The after-move-in service matrix in Figure 13.1 provides an example of an aggressive plan with many activities. You can use a similar approach to plan your after-move-in service activities and also as a reminder of services that you might add in the future.

As part of any expanded service program, pay close attention to the smallest details. Think in terms of showmanship. For instance, don't stop at offering a menu of maintenance services, make the menu more appealing with related paperwork that is professional and attractive. If your focus is social activities, ensure that you consider refreshments, parking, lighting, music, and similar details. Perhaps you will decide to expand warranty services to a two-year warranty for materials and workmanship instead of a one-year plan. Along with that extension of service, you might want to ask warranty reps doing inspections to arrive at the home wearing a company shirt or blouse and name badge. Whatever choices you make for after-move-in services will produce greater benefits if you leave no detail to chance.

Transitions: The Move-In Process

Moving is seldom named as anyone's favorite activity. The numerous details—some easily overlooked—give builders an opportunity to provide thoughtful (and usually inexpensive) services. Think of the chronology of moving: the preparation, the actual move, and the aftermath, as you look for ideas to transition your buyers smoothly into their new homes and gradually wean them off of the attention they have become used to.

Moving Preparation

Provide home buyers with useful moving preparation tips and tools. Labels with your company logo, packing tape, and markers could be packaged as a thoughtful gift. Your new home owners will appreciate cards that say, "We've moved to our new [Builder] home." Your buyers can each use these note cards to notify friends and relatives of the new address and phone number. Provide these cards as a surprise gift, a service that advertises your company at the same time.

No one likes moving. You are in a position to watch the process and come up with great ideas to help. Figure out what you can do to make moving easier, faster, and more fun. Thoughtful touches can impress the buyers you have and lead to earning new referrals.

FIGURE
13.1 After-Move-In Service Matrix

	Sales	Selections	Construction
Transition services	Visit new home owners and deliver designated community move-in gift within 2 weeks of settlement.	Host home care reception at sales center; provide attendees with *Home Care Basket* of sample products and merchant coupons.	On move-in day, install *Parking Reserved for New Home Owner* signs in front of home. Ensure that driveway and walks are clean. Visit new home owners during move-in, remind them that move-in materials and tools are available for their use and offer to collect flattened packing boxes upon their call.
Enhanced warranty	Welcome home owners with questions at sales center and assist them in reporting items to warranty staff as needed.	Communicate with warranty staff on potential details about home care that home owners need to know.	Welcome home owners with questions at the construction office and assist them in reporting items to warranty as needed.
Courtesy repairs			Assist warranty staff with Service Sweeps.
Information	Conduct Tax Benefits of Home Ownership Seminar between January and March.	Conduct Home Owner Workshops: • Closet Organizing • Wall Decor • Window Treatment	
Social activities	Host Private-Showing Reception for existing home owners at new communities prior to Grand Opening.	At closeout, host reception with silent auction of accessory items. Donate proceeds to charity.	Conduct summer barbeque for home owners, staff, and community trade contractors.
Feedback, repeat and referral buyers	Ask new buyers why/how they chose [Builder]; report results to chief service officer.	Review and discuss feedback; select action items and set implementation date.	Review and discuss feedback; select action items and set implementation date.

FIGURE
13.1 *Continued*

	Sales	Selections	Construction
	Review and discuss feedback; select action items and set implementation date.		

	Closing	Warranty	Chief Service Officer
Transition services	Mail *Packing Tips Kit* 60 days prior to anticipated settlement date. Call home buyer two weeks prior to settlement to review details.	On move-in day, check with home owner for questions and deliver Move-In Survival Kit.	Monitor and ensure support for all assigned activities.
Enhanced warranty	Send *Thank You for Buying* card on the anniversary of the settlement date.	Watch for items home owners did not notice and volunteer to correct them. Maintain a supply of minor consumable items to give to home owners. Leave a thank you coupon for coffee or ice cream at conclusion of repair visits.	Coordinate with purchasing staff and trades to identify items that might be needed and ensure ready supply for warranty to have on hand for home owners.
Courtesy repairs		Conduct Service Sweeps in spring and fall.	
Information		Conduct Home Owner Workshops: • Landscaping to Save Work and Water • Home Maintenance Tips • Recycling and Organic Home Care	Assist with seminars and workshops by establishing relationships with suppliers and potential instructors as needed.
Social activities			Assist with catering as needed.

Continued

FIGURE
13.1 Continued

	Closing	Warranty	Chief Service Officer
Feedback, repeat and referral buyers	Review and discuss feedback; select action items and set implementation date.	Review and discuss feedback; select action items and set implementation date.	Acknowledge and respond to comment cards and hot line contacts; follow up as needed. Ensure that survey summaries are circulated to all departments for discussion and action. Track and circulate rate of repeat and referral buyers.

Forestall last-minute stress with a pre-closing conference call two weeks prior to settlement to review tasks that need to be addressed before that appointment. The sales staff oversees this conference for some companies, and for others, the closing coordinator would manage this contact. Follow up with a confirming letter.

Move-In Day

At orientation, ask buyers when they will be moving in. Once you know this date, construction might place a "Reserved for New Home Owners" parking sign in front of the new home and ensure that walks and drives are clear of snow, ice, or mud. Assemble move-in aids that buyers can check out for a day or two, such as carpet runners and pads, hand trucks, dollies, and small hand tools. Perhaps you have a staff person available who can house sit by appointment to admit installation personnel or delivery people. (A cell phone or laptop allows this individual to continue being productive while performing a service for the home owner.)

A bouquet of flowers on the counter provides a welcome sight on move-in day. A new home owner also could use a move-in survival kit with trash bags, paper towels (or better yet, terrycloth towels with your company logo), toilet paper, a picture-hanging kit, beverages, and snacks. Think a step beyond the common paint touch-up kit. Home owners are accustomed to receiving small hand tools, cans of touch-up paint, and a tube of caulk; while those items are important, they generate only modest excitement. In the summertime, you might consider leaving your new home owner a small cooler stocked with "designer" water with a large bow and welcome note attached.

Plan to have someone check in with the buyers during their move. At that contact, simply ask how things are going and whether they have any questions. Just knowing that someone is available is reassuring and impresses many buyers. The clear message is "We are not going to disappear just because you've paid for the home."

To take this attention even a step further, provide a menu of move-in services that involve people and tools. These services might include connecting the ice maker, washer, and dryer; installing drapery rods, hanging pictures and mirrors, and later, touching up paint damaged during move-in. One company looks after landscaping for the entire first month after closing. Not only are the home owners impressed, the community looks terrific.

Moving Aftermath

As one superintendent observed, "Packing boxes are going to end up in my trash bins anyway, I may as well offer to pick them up and get credit for the service. At least that way home owners take time to flatten them, and I can send the boxes to be recycled." Letting home owners know this help is available relieves them of another moving annoyance.

Another after-move-in service might involve a visit from the salesperson, who delivers a welcome gift such as wine glasses or a coffee carafe with the company logo discreetly displayed. The after-move-in warranty meeting described in Chapter 3 offers another opportunity for positive contact. Or, as also mentioned earlier, the selections professional who worked with the home owner might visit to deliver a basket of cleaning-and-polishing supplies—and to see the end result of the choices the buyers made.

Providing a membership to the local recreation district, a subscription to the local paper, or similar items can be an important step to getting newcomers settled in. Double check that new home owners are aware of and set up for trash pickup and recycling services. Someone from the builder's office might call to confirm that they remembered to put utility services in their names—having the electricity shut off in the middle of unpacking is not the least bit amusing.

The on-site construction team attends to ongoing construction details such as parking for the trades and delivery vehicles, speed limits, trash, radio volume, language, start and stop times for work (especially those times that involve power tools), and keeping streets and vacant home sites free of trash. This attention to details impresses new home owners (and their visiting friends, who may be in the market for new homes, as well).

Enhanced Warranty Service

Some builders are following the trend of the automotive industry by extending traditional warranty coverage for longer periods. For instance, several companies have increased the typical one-year warranty to two years' coverage of materials and workmanship. They secured the cooperation of their trades and suppliers before advertising this service. Others give home buyers a choice among two or more levels of protection: the traditional coverage at no cost or upgraded coverage at a cost. This optional deluxe warranty coverage might last longer, carry stricter guidelines, or both. To establish reasonable costs for such programs, builders need accurate historical repair-and-expense data.

You might increase customer goodwill simply by revisiting warranty exclusions and other policy choices. For instance, you might make normal warranty coverage transferable (or assignable) to subsequent buyers. Reconsider as well items that are totally excluded from warranty coverage. By negotiating supplier agreements accordingly, sod and shrubs can be warranted for one growing season or even a calendar year.

Frequently, warranties completely exclude coverage for nonstructural concrete flatwork (long the bane of builders worldwide). In reality, this exclusion effectively eliminates any supplier, installer, or supervisory accountability—a practice that understandably seems outrageous to people spending thousands of dollars for a new home. Some builders do cover concrete flatwork; many do not because they got tired of arguing about the type of repair they provided for cracks. Because home owners always want replacement, covering it may not be your solution, but providing nothing is equally inappropriate. **Consult your attorney before making any of these or other changes in your warranty document.**

Along similar lines, review the section under the subhead, "Cosmetic Damage" in Chapter 5. Consider whether more lenient "back door guidelines" could increase home owner satisfaction and result in more sales. Likewise, review policies and timetables for so-called courtesy repairs such as drywall separation and nail pops, touch-up caulking and grouting, stucco separation repairs, and addressing settling in foundation backfill areas. As with any new service procedure, evaluate the return on your investment before being more generous with such services (and by all means consult with your attorney).

Next, double-check that the repair appointment suggestions in Chapter 6 are familiar to your technicians. Ensure that daily interactions between home owners and repair personnel contribute positively to your company's reputation. To add a touch of showmanship to ordinary warranty repair appointments, warranty techs might arrive in pairs, with cleaning supplies and tools at hand, prepared to leave their work area spotless. In-house warranty technicians should correct listed items and check for others that may have been overlooked. Your trades' personnel can do the same thing.

Offering warranty service appointments in the evenings or on weekends is a challenge for several reasons. However, if those challenges can be met, this kind of flexibility probably will be extremely popular with busy home owners.

Real estate agents often look to the builder for help with items when they list a home. If this situation commonly occurs in your region, you might organize a Broker Preparation Package starting with a warranty inspection, appropriate work orders, and guidance on maintenance items. Such a program could blend easily with maintenance service menus discussed in the next section.

Maintenance Services

For maintenance tasks, builders can choose from many levels and styles of services. If you determine that extra maintenance services would appeal to your market, match the level and style to your service capacity and needs of your home owners.

Maintenance Items Addressed During Routine Warranty Repair Visits

At the simplest level, you can incorporate a modest number of maintenance services into routine warranty repair visits. At the same time that your technician adjusts a door and touches up backsplash caulking, he or she might replace the furnace filter, smoke detector batteries, or a difficult-to-reach light bulb. Maintaining a supply of cabinet door bumper pads, doorstops, hard-to-find styles or sizes of light bulbs, and other parts allows technicians to solve problems for home owners on the spot at a relatively low cost to the company.

Consider several factors before offering off-hours service. For one, a significant portion of repairs require daylight for proper execution, including drywall, paint, and exterior work of almost any type. Also, independent trade contractors who help you build homes may find the concept of evenings and weekends amusing, and the few repairs that could be performed in off-hours could fail to eliminate the need to still have repair appointments during normal hours. Administrative staff and supervisors would need to be available to answer questions. Having some personnel work extended hours could mean being short of staff during normal business hours. Finally, you could experience a significant impact on wages and salaries from adding more personnel or compensating existing personnel for working nontraditional hours. All of these issues have potential answers; be certain you have those answers before proceeding.

Service Sweeps

In a more aggressive approach, you can offer service sweeps at no cost or for a small fee. Send invitations to all home owners in a community and describe the service to be provided to those who RSVP. These invitations will earn the best response if you send them two to three weeks prior to the service-sweep date, and you schedule the work on a Saturday. Select the services your team will perform carefully. Each quarter you could offer something different. Trade contractors could participate. For instance, arrange for your heating contractor to perform an annual inspection and adjustment of the heating, ventilation, and air-conditioning (HVAC) systems. Offer window cleaning, yard treatment, or a package involving filters, batteries, lightbulbs, door adjustments, and so on.

Menu of Services for a Fee

Figure 13.2 lists many potential services that might be included on a standard menu of after-move-in home maintenance tasks. Warranty personnel provide some of these services during the term of the warranty, and the services become

FIGURE

13.2 Maintenance Menu Ideas

Item	Price	Item	Price
Air-Conditioning Clean or replace filters. Have annual inspection done. If necessary, have it charged with coolant.		**Electrical** Reset tripped breaker. Replace bulbs in high fixtures. Test smoke detectors and outlets with ground fault circuit interrupters (GFCI).	
Bathrooms Check caulk; touch up or replace as needed. Check grout; touch up or replace as needed.		**Exterior Brick/Stucco Veneer** Repair cracks.	
		Faucets Clean aerators. Replace washers/o-rings.	
Cabinets Tighten hinges. Adjust to level.		**Fireplace** Clean chimney. Inspect damper for correct operation.	
Decks Apply water repellent. Replace warped boards.			
		Fire Extinguisher Do the annual inspection.	
Disposals Unclog.		**Floors** Seal concrete floor. Polish marble. Clean, seal, and/or touch up grout. Stretch carpet. Re-coat hardwood flooring.	
Doors Refinish exterior door. Adjust weather stripping and/or threshold. Adjust interior door.			
		Foundation Fill cracks with epoxy. Correct settlement of backfill.	
Drains Unclog drain. Unclog toilet.			
		Glass Replace broken windows.	
Driveways, Walks, Steps Seal minor cracks.			
Dryer Clean dryer exhaust vent.		**Gutters and Downspouts** Clean and check.	

FIGURE
13.2 *Continued*

Item	Price	Item	Price
Heating systems Have annual inspection done. Clean or replace filter. Clean humidifier. Relight pilot.		**Smoke Detectors** Test them. Replace batteries.	
Holidays Install decorative lighting.		**Termites** Inspect foundation and base- ment, if any. Have exterminator treat any termites.	
Landscaping Fill depressions. Replace dead or sick shrubs and/or trees.		**Trim and/or Moldings** Re-caulk as needed.	
Plumbing Check drain traps.		**Walls and/or Ceilings** Repair minor cracks. Paint as needed for repairs.	
Security Systems Replace batteries.		**Water Heaters** Wrap with insulation. Drain sediment. Check temperature. Check pressure valve.	
Septic System Order annual inspection. Have system cleaned.			
Shower Clean shower heads. Re-grout as needed. Re-caulk as needed.			

more useful to home owners once the house is out of warranty and the home owner has become accustomed to them. You would almost always offer these services for a fee. Busy or older home owners and those home owners not interested in home maintenance often appreciate having people they know come to their homes to assist with such items. You can price the services in packages or a la carte.

Property Management Services

In some communities, builders can offer a complete range of property management services to investor-owned homes and those occupied by part-time owners or renters. To be profitable, builders need to verify that the volume of such homes will support the service. This type of service blends well with typical warranty processes and systems. Your version might cover interior physical repairs alone, or it might extend only to exterior care items (such as pools and landscaping). Depending on the numbers involved, you could expand your property management to include leasing services; some builders even set up a real estate office to handle resales for their home owners.

After-Move-In Selections

Depending on your price point, you may have a significant number of home owners who eliminated desired selections from their initial wish list to stay within their budgets. Once they have moved in, provide opportunities for them to purchase these selections after closing. This service can generate additional profit. Promoting such a program with home buyers during their selection appointments influences their initial choices. They may order items that cannot be installed later (spiral staircase) and save others for later (covered patio).

The list of potential items for an after-move-in selection program could include interior built-ins or trim, ceiling fans, garage door openers, patios, patio covers, decks, rock or tie walls, fences, garage workshops, and basement finishes. Remember to include in your estimates and to charge for the time and costs involved for items that require home owner association approval, building permits, and warranty coverage.

Informational Support

Naturally, if a home owner calls to ask for a paint color number, your personnel will provide that cheerfully. All builders can set up timely and appropriate reminders with postcards or e-mail.

In addition, depending on the climate, you can remind them about changing or cleaning filters, removing hoses from exterior faucets prior to the first freeze, and operating their air conditioners before scorching weather hits.

Seasonal newsletters that include maintenance reminders such as those listed in Figure 13.3 serve as a logical communication tool for maintenance tips and announcements. Add community announcements of general interest, notes welcoming newcomers, and other community news. More and more companies are expanding this kind of correspondence to keep their names in front of customers in a positive manner. In fact, some builders are sending newsletters for 10 years instead of the usual one or two. Soliciting referrals has no time limit.

More elaborate programs for providing information offer mid- and large-volume builders an opportunity for another layer of services. The tips for organizing home buyer seminars in Chapter 1 can easily be applied to after-move-in home owner educational programs. Whether you present them as lectures, demonstrations, or hands-on workshops, possible topics include a wide range of possibilities:

FIGURE

13.3 Seasonal Maintenance Suggestions for Newsletters or E-Mail Reminders

Monthly (or Follow Manufacturer's Recommendations)

- Clean and test smoke alarms.
- Test and reset breakers for ground fault circuit interrupters (GFCI).
- Change or clean furnace filter.
- Drain sediment from water heater per manufacturer's instructions (local water quality determines needed frequency).

Spring

- Check and operate air-conditioning system.
- Adjust registers and confirm that cold air returns are clear of furniture or draperies.
- Make certain the air-conditioner compressor is level and clear of debris.
- Turn the humidifier off.
- If your home has a private well, have the water tested.
- Start and adjust sprinkler system. Test exterior faucets for broken pipes.
- Check garage overhead door, tighten bolts as needed, and lubricate springs with motor oil. Have other repairs done by professionals.
- Clean gutters and confirm that downspouts or splash blocks drain away from the house.
- Look for settling of backfill soils and fill in where needed.
- Check exterior caulking and touch up.
- Check exterior paint and stain surfaces (especially stained doors) and refinish as needed.
- Inspect grout around tile (floor or wall) and touch up.
- Wash windows and screens, clean weep holes, and lubricate tracks.
- Inspect for shrinkage damage such as minor drywall cracks and separations of wood trim. Repair as needed.
- Plan your first barbecue.

Summer

- Regularly check sprinkler head adjustments.
- Check interior caulking and touch up.
- Inspect grout around tile (floor or wall) and touch up.
- Pour a quart of water down the basement floor drain. As water in this drain evaporates, sewer odor can seep into the house.

Fall

- Operate (test) the heating system.
- Adjust registers and confirm that cold air returns are clear of furniture or draperies.
- Clean the humidifier per manufacturer's instructions.

Continued

FIGURE
13.3 *Continued*

Fall (Continued)

- Adjust or replace weather stripping on exterior doors as needed.
- Check the fit of exterior doors at their thresholds. Many designs are adjustable. Use a quarter to turn the large screws along the top edge.
- Drain the sprinkler system.
- Remove hoses from exterior faucets. Even "freeze proof" faucets end up with a broken water line if the water in the hose freezes and expands into the pipe.
- Inspect chimney for nests.
- Review safe fireplace operation. Chimneys for fireplaces and wood-burning stoves need professional cleaning at regular intervals.
- Check overhead door of garage, tighten bolts as needed, and lubricate springs with motor oil. Have other repairs done by professionals.
- Clean gutters, check downspouts, and confirm that splash blocks drain away from the house.
- Check foundation, concrete slabs, and yard for settling of backfill soils, fill in as needed to maintain positive drainage.
- Check exterior caulking and touch up.
- Wash windows and screens; lubricate tracks.

Winter

- Follow all instructions for safe operation of any fireplace or wood-burning stove.
- Brush snow off gutters and away from downspouts.
- Remove ice and snow from concrete surfaces as soon as possible.
- Avoid using de-icing agents with damaging salts.
- Pour a quart of water down the basement floor drain. As water in the drain evaporates, sewer odor can seep into the house.
- On pleasant days, open windows to allow the house to breathe.
- Decorate safely for the holidays. Do not overload circuits or use worn extension cords.

- Routine maintenance tasks such as draining water heaters, correcting nail pops and minor drywall separations, replacing caulk or grout
- Organic home care, including conservation tips, recycling techniques, recipes for green cleaning products
- Interior design projects such as faux painting, wallpapering, window-covering ideas, mirror and picture-hanging hints, and lighting tips
- Landscaping techniques that minimize water usage or help to create a beautiful setting for the new home
- Closet-organizing techniques, tools, and products

⊛ Tax benefits of home ownership, a topic especially appropriate for first-time buyers

> Regardless of the topic, you might want to clarify that attendees are welcome to bring a friend—the more people who know how terrific your company is, the better.

Social Activities

From something as passive as organizing a community directory (names included with permission of the home owners) to regularly scheduled activities, social events can contribute significantly to your after-move-in service program.

Combining good food with interesting activities provides a great way to gather a crowd and stimulate conversation. Some builders host quarterly or semi-annual "Meet Your Neighbor" receptions. As always, invite folks to bring a friend, and have sales staff available to show models to interested prospects. This reception might blend with efforts to attract volunteer ambassadors (see Home Owner Ambassador Program later in this chapter) for your building company. High-end builders might consider hosting a catered house warming for their home owners.

Events might be organized around seasons or holidays. One builder offered his models as a setting for weddings on Valentine's Day. Golf tournaments, art shows, out-to-lunch groups, and children's activities all offer opportunities to stimulate positive word of mouth. And don't overlook community involvement. One attention-getting means of contributing involves setting up a silent auction of model home furnishings at community close out, then donating the proceeds to charity or a local cause. In addition, by announcing the donation in a press release, you will also gain word-of-mouth publicity from the home owners who buy the furnishings because they will tell their friends about the event and your company.

Feedback

The services you select may need to be revamped over time. You also will need to monitor the success of the traditional warranty service elements of your after-move-in program. Stay in touch with your home owners so they can easily and conveniently share their impressions and ideas with you at any time in the formats they prefer. Establish a flexible and ubiquitous net to gather feedback without being intrusive, repetitive, or annoying. This survey approach means implementing more than one method, including some passive (the home buyer volunteers comments) and some active (you ask for comments).

Comment Cards

Provide self-addressed, postage-paid comment cards as part of all routine warranty contacts and special events. Include the same form in your home owner guide and on your Web site's home page. Acknowledge and respond appropriately to all comments you receive.

Customer Hotline

The goal is to keep customer communication open and provide a fast track to assistance. Consider a phone and e-mail "hotline." Again, acknowledge receipt of the contact immediately and follow up appropriately.

Written/Phone Questionnaires

One or two periodic, comprehensive surveys are the foundation of broad-based insights. To use them well, follow the suggestions listed below:

* Survey shortly (two to four weeks) after closing, rather than at the closing table when customers' emotions are distorted and magnified by stress, excitement, and exhaustion. Ask about early steps in the process as well as the condition of the home upon delivery.
* After a home owner has been in the home about a year, ask about floor plan, quality, normal warranty services, and the extra services you offer.
* The following two questions are the heart and soul of your surveys: "Would you buy again? Would you recommend our company?"

Manager Interviews

Consider implementing in-depth management interviews to expand upon statistical data that accrues from survey responses. Once or twice a month, each manager interviews a home owner in his or her home. These visits usually take place in the evening and last approximately 30 to 45 minutes. Simply ask, "What was it like to do business with [Builder]?" Take brief notes during the interview and add details immediately after the meeting, while your memory is fresh. The benefits of interviews are significant. You will find some listed below:

* Having a vice president or manager sit in a home buyer's living room makes an impression. Home owners feel important and listened to.
* Customers often make excellent suggestions. They are in a position to know what would make the process more enjoyable.
* When the staff and the trades know that managers visit home owners to get their feedback, service goals take on new significance.
* Managers stay in touch with the realities that frontline people face and fine tune their customer service insights.

Managers discuss the results and add their insights to the statistics that summarize written surveys and other feedback results. The goal is to ensure that the company is identifying appropriate action items, that nothing is overlooked, and that survey statistics are being interpreted correctly.

Inner Circle

Perhaps you would prefer to create an "inner circle" of home buyers, a group of customers who agree to provide detailed feedback. In essence, you invite several actual buyers to "shop" your after-move-in processes. If possible, their identity should be kept confidential so you learn about how the process works normally, rather than when everyone is making a special effort to impress an inner-circle buyer.

Be candid with participants regarding the time commitment involved. Show appreciation with a substantial thank you. For example, provide a gift certificate for furnishings or window coverings. This group reports to you on company successes and failures at regular intervals. Interview participants by phone or in person and, before disbanding, host a group discussion with all inner-circle participants invited.

Unstructured Feedback

Daily interactions with frontline personnel offer tremendous insight into home owner opinions so monitor these less formal communication systems. Stay alert for great ideas and watch for patterns and trends. Warranty staff can note recurring questions or issues as they work with home owners. Anecdotes, when compared and discussed, can call attention to opportunities that other feedback methods might overlook.

Home Owner Council

Separate and distinct from any home owner association function, this group of volunteers acts as a conduit for ideas—their own and those of their neighbors. They meet with company representatives quarterly, or more often if needed, to discuss both positives and negatives.

Building Sales Through Service Excellence

Keep in mind that all of the previously mentioned efforts have the same objective: increasing repeat and referral sales. If you don't already know, discover your current rate of repeat and referral buyers and establish an ongoing system to track those statistics. Set up a target and report progress to company personnel along with customer feedback.

Home Owner Referrals

Some customers think of referring others, but many do not. They need to be prompted. To obtain their referrals, create a system or program that makes referring a potential home buyer easy. Keep referral cards readily available at all business locations, in your home owner guide, and on your Web site. Save a corner of the community newsletter to ask for referrals.

Referral Rehearsal. When a home owner says yes to the question, "Would you recommend our company?" take time to ask exactly what he or she would say about your company. This information is valuable to you, and the rehearsal makes it more likely that your home owner will repeat the remarks to others. Their quotes may be perfect for a testimonial in your print ads as well (remember to obtain written permission to use such comments).

Training. Train company personnel—everyone, not just salespeople—in how to ask for referrals. Staff members who do not see themselves in the role of salespeople may find asking for referrals uncomfortable. But by showing them how easy and straightforward such conversations can be, you help them get over that discomfort.

First, set the scene, for instance suggest to staff that a warranty technician has just completed a work order and after asking the home owner to sign it, the home owner says, "Thanks for taking care of this so quickly, we appreciate it." Ask your staff person to respond with something like "If you are pleased with our service, will you refer your friends to us?"

Perhaps the superintendent for a small-volume builder stops by a few weeks after new home owners move in. She asks how the move is going and hears a rave review about the home and the usual comments about moving. Be certain the superintendent is comfortable responding with a referral request such as, "We appreciate your business and look forward to doing business with you again. I also would like to help any of your friends who are in the market for a home."

Or perhaps a warranty manager concludes an inspection on a short list of items. When the home owner observes that this is a great quality home, the manager should be able to follow up with "I'm glad to hear that. We've built our business on customer satisfaction. How would you rate our service to you? <assuming more positive comments come from the home owner> If you have friends in the market for a new home, would you feel comfortable referring us to them?"

Referral Appreciation. When you receive a referral or recommendation, simply expressing sincere appreciation may be enough, or you might prefer to give a modest token of thanks. Consider providing the same gift to the *referrer* and the *referee*. Avoid excessive payments, which might suggest to some people that you must buy referrals. Customers refer others when they genuinely believe their relative, friend, or acquaintance will be well served, not because of the promised two-week vacation trip.

Employee and Trade Contractor Purchase Programs

Depending on your product price point, your own employees and trade contractors might offer an untapped source of sales. These people know your service efforts from the inside out; their good opinions are a clear sign you are doing many steps right. Consider programs that stimulate purchases from these groups. Besides creating more sales, salespeople take pride in mentioning to prospects that many employees and trades' personnel own the builder's homes. Such an expression of confidence impresses most buyers.

Home Owner Ambassador Program

A technique that is particularly effective in active adult communities involves creating a team of home owner volunteers who act as hosts and hostesses, especially on weekends and other high traffic days. Such a program can be a terrific way to support your salespeople. The credibility of enthusiastic veteran home owners makes a strong impression on prospective buyers.

Home Owner References

Besides displaying complimentary letters from home owners in a scrapbook, builders can benefit from compiling a list of home owners who are willing to talk to interested prospects. Small-volume and custom builders in particular find this approach helpful because they often do not have model homes. Therefore, set up an appointment for a prospective buyer to visit one or more existing home owners, tour their homes, and hear their recommendations. This practice can be a vital part of a custom builder's marketing efforts.

The service horizon possibilities discussed in this chapter merely scratch the surface of what builders can do to preserve and extend good relationships with their home buyers. With imagination and attention to details, a strong program evolves to the benefit of the home owners and the builder. The process begins with seeing potential for improvement or expansion and thinking differently about service.

Staffing for After-Move-In Service

CHAPTER

14

Along with the goal of exceeding customer expectations come new staffing questions. If you want to increase home owner satisfaction and loyalty by providing nontraditional after-move-in services, who will plan and manage those activities? If your company decides that, at least for now, it will concentrate on developing service excellence in warranty, what configuration of warranty staff offers your company the best opportunity for success? To answer these questions, begin by examining the changing nature of service in the home building industry.

Changing Perspective: An Equal Voice

Until recently, few companies of any size had one person overseeing the entire customer experience—organizing communication, paperwork, and processes into a cohesive and enjoyable whole for the home owner. Now this position is more common, as builders increasingly measure and track the positive sales impact of healthy long-term relationships with home owners. You do not saunter accidentally into such a service reputation. You need clear service goals, a plan to achieve them, and someone whose primary assignment is managing that plan.

Companies of all sizes are acknowledging that successfully managing the customers' experience requires focused attention and coordination that is unlikely to occur when the steps in the process are managed by separate departments (in large companies) or individuals (in small firms). Giving customers a voice equal to those of other department heads can be the catalyst that generates customer-centric operations. This job formalizes the connection of each department to the company's customer satisfaction goals.

The person who holds this job will be most effective when his or her title is equal to those of other department heads. Organizational charts often include Chief Executive Officer (CEO) or Chief Financial Officer (CFO). Those companies can create a Chief Service Officer (CSO). If your company has vice presidents (for instance, Land Development, Operations, Sales and Marketing), a title such as Vice President of Customer Relations or Customer Service or the Customer Experience) will work. If your company refers to department heads as directors, then Customer Relations Director is fine. Likewise, if manager is the typical department head title, Customer Relations Manager will work. In small-volume companies, the assignment is typically managed by the owner with the support of administrative and construction personnel. Figure 14.1 offers a job description for this assignment, by whatever title.

This job is somewhat unique in that the responsibilities cut across all departments and functions. The individual who fills the position should attend all strategic planning sessions and management retreats as well as routine management meetings. The company owner should establish how much authority or influence this position will have over various departments.

Anticipate that some veterans may need time to adjust to sharing authority with another professional. An ability to build healthy working relationships will be critical to success. The person who takes on this role has an opportunity to accomplish exciting goals for the company and its home buyers. To do so requires ingenuity, organizational skills, and above all the ability to work successfully with a variety of personalities and egos, beginning with his or her closest colleagues.

Service Culture

Another profound shift toward fostering customer satisfaction is occurring in the human resources aspects of home building. In the past, many builders have assumed that, of course, all employees know that customer service is important and is part of their jobs. Now many more of them know they must lead their companies, organize their businesses, and train their employees toward this customer service ethic.

Service DNA

Service, customer satisfaction, quality, and measurable improvements are appearing in job descriptions and performance reviews. This change builds these desirable elements into the DNA of a company.

Many staff meeting agendas list customer satisfaction as the first topic of discussion. Recognition and rewards for service excellence have become as important a part of company meetings as sales awards.

Universal Service Guidelines (USG)

In most companies, a wide range of habits for managing phone calls, e-mail, and meetings exist. But these everyday-everyone behaviors can profoundly impact a company's reputation. If an organization lacks common performance guidelines, each individual's personal values are inevitably applied—with wildly fluctuating results.

Savvy builders are addressing the fundamental issue of the service that their trades, associates, and company employees and departments can expect from each other. They no longer leave that service to chance. In the past, only a few builders defined service. Others didn't plan carefully enough or left service issues to manage themselves. Consequently, that practice meant that the expectations

FIGURE

14.1 Sample Job Description: Chief Service Officer

Position Title: Chief Service Officer

Purpose: Plan and coordinate systems and training to achieve desired customer satisfaction levels and sustain company focus on all aspects of customer service.

Duties and Responsibilities:

- Develop and implement procedures that ensure customers enjoy an exceptional new home building experience and that result in high customer satisfaction ratings and increased repeat and referral sales.
- Oversee development, implementation, coordination, and maintenance of Universal Service Guidelines (USG) to ensure that all employees, home buyers, and associates receive consistent and exceptional service at all times from all departments.
- Apply the USG in all dealings with home buyers, colleagues, or associates.
- Monitor written materials customers encounter and recommend changes to ensure forthright and hospitable communication to customers.
- Select and institute a portfolio of feedback systems to identify improvement opportunities and recognize accomplishments.
- Circulate feedback results to appropriate personnel on a regular basis; ensure review, interpretation, and where needed, action plans for improvement and innovation in all aspects of customer services.

Sales and Marketing

- Orient and update sales personnel regarding service procedures, including pre-construction conference, frame stage tour, orientation, and warranty service.
- Assist sales staff in responding promptly and accurately to buyers' questions about product, procedures, and services.
- Provide sales offices with [Builder] Home Owner Guides and other materials to promote positive and realistic customer expectations regarding the product, the process, and related services.
- Assist in planning and managing programs to increase repeat and referral sales.

Construction Process

- Develop procedures for change orders that balance sales and marketing's need for flexibility with construction's need for orderly processes and adequate notice.
- Organize and implement preconstruction conferences and frame tours for home buyers to keep them informed and involved in the process, answer their questions, and resolve issues promptly.
- In large communities, develop and implement home buyer seminars.
- Work with sales and construction to establish site-visit policies that ensure safety, customer involvement, and orderly operations for on-site personnel.
- Assist sales and construction personnel in resolving home buyer issues during construction, documenting the results, and identifying needed changes to prevent recurrences.

Continued

FIGURE
14.1 *Continued*

Delivery

- Train personnel to perform home owner orientations in a consistent and positive manner.
- Monitor completion of orientation items and ensure timely response by construction personnel.
- Develop and maintain a regular reporting system that summarizes the number, nature, and completion of orientation items.
- Assist in identifying recurring items; work with construction and trades to eliminate them, where possible, to improve product quality.

Warranty

- Create procedures for processing routine, emergency, and out-of-warranty items.
- Plan and oversee customer service training for warranty staff.
- Appraise performance of each warranty staff member.
- Maintain a reporting system summarizing the number, nature, and completion of warranty items.
- Identify recurring items; work with construction and trades to eliminate them where possible.
- Document and appraise performance of trade contractors in warranty service and treatment of customers. Work with the [builder]'s trades to improve customer service performance and attitudes.
- Control costs by efficient use of personnel, appropriate back charges to trades, and feedback to construction and purchasing regarding recurring items.

Supervisory Responsibilities: Warranty Manager

Reports to: President

of those groups depended on each individual's imagination while the performance of this service was left to the discretion of the individual employees or departments.

Consider what would happen if everyone within your organization agreed to Universal Service Guidelines designed to serve all of the company's customers well: home buyers, trade contractors, associates, and employee/department customers. Not only could home owner Jones expect an exceptional level of service, so could the plumber, the selection center, construction, purchasing, and you—all the time—from your colleagues.

An example of Universal Service Guidelines appears in Figure 14.2. Begin with this sample and invite all company personnel to offer additional points or adjustments. Participating in developing this service code will foster commitment and create a sense of ownership. The Chief Service Officer (CSO) oversees the creation of these guidelines. In a small-volume building company, the

FIGURE

14.2 Universal Service Guidelines

Communication

- Set expectations with your customers by informing them in writing as to the best way for them to reach you (day, time, and method) and likewise learn the best way for you to reach them.
- Create an acknowledgment expectation of one business day and a personal standard to acknowledge within four hours.
- When you are unable to resolve an issue in one contact, estimate the time needed to do so and commit to a follow-up date.
- Record this date and keep the appointment even if you are still working on a final answer: "I'm just calling to tell you I have nothing to tell you." Set up the next appointment.
- Create a separate phone or communication log for each contact. This log allows you to sort, prioritize, delegate, make notes, and file contacts. Document the date, time, and points you made. Standardize abbreviations you use to make this task go faster.
- Except for real emergencies, batch communications whenever possible. For example, minimally return calls once in the morning and once in the afternoon.
- Set up talk-to agendas for the people you communicate with regularly. Accumulate a list of subjects and cover them in one call to each person. They will all appreciate not being interrupted repeatedly, and you will save yourself time.
- When you are unavailable for an extended period of time, arrange to have someone else cover for you—at least to the extent of acknowledging your calls and explaining when you will return.
- Avoid making commitments for others without their consent. "I'll have him call you this afternoon" sets up the customer for disappointment. "I'll give him the message" is more accurate unless you know for certain the party you are referring to is available and willing to follow up.
- When a customer comes to you with a problem, own the issue. While the problem may be outside of your authority or expertise, ensure that the party who should handle it learns about it and follow up with the customer to confirm that the issue was resolved. Instead of saying, "You'll have to call Joe on that," talk to Joe yourself and ask him to contact the customer. "It's not my department" fails to satisfy customers.
- Plan points to make in each call. Experience shows that planned calls average 7 minutes; unplanned calls take 11 to12 minutes.
- When you place a call be ready to leave a message; research shows that 88% of the time the person you call will be unavailable; 50% of the time, the two of you do not need to talk directly anyway. For example, avoid saying just "call me" and instead say "I have two questions. First, . . . and second, . . ."
- In all phone messages, include the company name, your first and *last* name, phone *and* extension number, and e-mail address.
- If you need a call back, say so and provide a couple of times you will be available—then be available at those times.
- Follow up each contact immediately, mark your calendar and so on, before you move to the next task.

Continued

FIGURE
14.2 *Continued*

E-Mails

- Confirm that complete and accurate contact information automatically appears at the bottom of every e-mail: your phone and extension number, fax number, and mailing address.
- Avoid using your in-box as a staging area. Establish "save" or "print" folders for e-mails you want to follow up on.
- In e-mails, avoid the use of ALL CAPS, the typographical equivalent of shouting.
- Compose complex or emotionally charged e-mails off-line, edit the contents, and then paste them for sending.
- Keep in mind that you may be required to present all e-mails in a courtroom.

Faxes

- Include your contact information clearly and accurately on a fax cover sheet along with the total number of pages you are faxing.
- Avoid assuming that, because you faxed an item, it was received. Leave a phone message that the fax is on the way.
- When you are sending a long fax, call first to alert the receiver.

Meetings

- Meetings typically convene for the purpose of sharing or creating information or for making decisions. Before scheduling any meeting, ask yourself whether an alternate method of communicating would work as well.
- When meetings are necessary, make them productive and efficient. Include those people who can contribute, those who would benefit from hearing the discussion, and those affected by the outcome.

  People leading company meetings should take the following steps:

  - Set a date and time and provide sufficient notice so participants can prepare and attend without rescheduling other appointments.
  - Confirm that the location is convenient to the attendees and appropriate for the size of the group and length of the meeting.
  - Announce the meeting on one page (electronic or hard copy) and include the date, time, location, participants, and questions to be addressed. (If necessary, include directions.)
  - Schedule routine meetings at appropriate intervals—far enough apart that you have significant new information to cover, but not so far apart that people are making decisions in an information vacuum.
  - Create an agenda. For routine meetings use standardized agendas that include room for notes and an opportunity to raise new issues.
  - Remind the receptionist about the meeting, who will attend, what time you will take a break, and when you will end the meeting.
  - Bring sufficient copies of materials to be distributed.
  - Start on time.
  - Ensure the group proceeds through the agenda in an orderly fashion.
  - Encourage participation; allow everyone an opportunity to talk.

FIGURE

14.2 *Continued*

- Review notes from the previous meeting, as appropriate, and ask for relevant updates.
- As the meeting participants identify action items, make note of who will follow through and when an update is due.
- End on time.
- Distribute the minutes of the meeting within 24 hours.

Meeting participants should follow the suggestions listed below:

- Arrive on time.
- Arrive prepared.
- Turn phones, pagers, and radios off (unless someone is waiting for urgent information).
- Listen carefully.
- Contribute relevant, positive comments.
- Avoid sharing war stories unless they are relevant to the topic under discussion.
- Avoid negativism, whining, and blaming.
- Offer suggestions and solutions.

owner would oversee this activity. Once the suggestions are in hand, add these new service responsibilities to all job descriptions and performance review forms. Everyone in your company has customers and needs to practice service excellence.

Warranty Excellence First

No company can escape the fact that nontraditional after-move-in services produce better results when traditional warranty tasks are well managed including the following elements:

- A warranty staff structure appropriate to company volume and product locations
- Adequate number of field staff for in-house repairs and administrative support
- Matching the physical location of warranty operations to geographic factors where appropriate
- Personnel attitudes that turn company values into tangible acts of service
- A comprehensive plan for ongoing training and development

Department Structure

No universally right way to staff and organize a warranty department exists. Your search is for the staff level and structure that is appropriate to your company's

Personnel in small-volume companies face the challenge of being effective while they are pulled in several directions by multiple priorities and responsibilities. For example, the same individual may manage warranty service, bill paying, and permitting, but performance excellence remains a paramount goal, home buyers' expectations do not drop because they purchase their home from a small company.

situation. To determine what that is for your firm, keep in mind that daily warranty work must address the following three kinds of activities:

- Management. Meet with home owners, make decisions, supervise other personnel, and oversee warranty trade relations.
- Administrative. Document warranty work, follow up to ensure closure and customer satisfaction, create and update reports, and file documents, reports, and correspondence (electronic and/or hard copy, as appropriate).
- Technical. Perform repairs, installations of backordered items, and the like.

Technology can support all of these activities and allow fewer people to accomplish more in less time.

Small Volume: One Person with Several Roles. For small- to mid-size companies, having a warranty manager who also carries tools to perform repairs can work well. This person may need administrative support in the office. In another arrangement, one person inspects items, performs administrative tasks, and is supported by a technician (who may also work part-time for construction doing quality control checklist items). This person would involve trade contractors as needed.

Mid-Size: Superintendents in Charge of Warranty. Many builders choose to have superintendents manage their own warranty work. This approach offers both advantages and disadvantages. Some advantages follow:

- Face-to-face accountability provides an incentive for the superintendent to build the home right the first time.
- Having supervised the building of the home, the superintendent knows it best, including the buyers' selections, change orders, and issues the buyers raised during construction.
- The superintendent typically has first claim on the time of trade contractors and, therefore, can best set priorities between new production and warranty repairs.

However, the disadvantages of this approach may be overlooked by company owners in the habit of having superintendents manage warranty. These disadvantages include the following:

- Time spent dealing with warranty issues is time not spent building new homes. Many builders who expected that dealing with home owners' warranty issues would train their superintendents to build better homes are still waiting for those results. Worse, their home owners are still having warranty issues.

- Conducting warranty inspections means the superintendent needs a full range of customer communication skills. Some superintendents may be unable to develop these skills; others may not want to.
- The knowledge and skill set necessary for handling warranty service differs from the knowledge and skill set needed to build homes. Asking one person to succeed at both assignments is expecting a lot—maybe too much.
- Superintendents are accustomed to handling trade items in person or by phone. If they manage warranty items the same way, documentation may suffer as a result.
- Progress in improving the quality of future homes may slow down if the field staff neglects to report issues, whether from lack of accurate documentation or for self-protection.

Warranty Manager. When production levels reach 45 to 50 homes per year, the superintendent approach usually begins to generate problems. Construction duties demand every minute of the day, and the superintendents have less time, patience, and diplomacy for warranty issues. One person needs to be in charge of and focused on warranty issues for the benefit of the home owners and the company. Figure 14.3 shows a job description for a warranty manager.

Experience shows that the person in charge should report to the company's executive officer, president, or owner. Rare exceptions exist, but having the warranty manager report to the head of construction or sales typically leads to conflicts of interest in which warranty generally loses. Genuine improvement in the quality of the homes requires objective checks and balances.

Warranty Service and the Chief Service Officer. If you divide after-move-in services between a warranty manager and another position, the two people must develop a respectful working relationship to succeed. They must share a common vision and strive for agreed-upon objectives. Occasional conflicts may prove inevitable; anticipate and work through them on a case-by-case basis. For instance, suppose the warranty manager must respond to a request for a new driveway from the most influential home owner in a community the day before the "Meet Your Neighbor Barbeque"?

Number of Personnel for After-Move-In Services

If a company builds three homes per year, one person may be able to handle warranty and some other functions as well—perhaps including some after-move-in extras. As volume increases, all

The overriding considerations in your decisions about staffing for warranty service, first, must be the comfort and convenience of the home owners; second, the ongoing improvement of the product; and third, warranty staff efficiency. These goals are best accomplished when one person focuses on warranty service responsibilities and oversees an appropriately sized team to manage those responsibilities.

FIGURE
14.3 Sample Job Description: Warranty Manager

Position Title: Warranty Manager

General Purpose: Establish and maintain staff and processes to address warranty service in accordance with legal requirements and [Builder] philosophy, foster a positive company image, and identify opportunities for improvement in company product, processes, and staff training.

Duties and Responsibilities:

Sales and Marketing

- Orient and update sales personnel regarding warranty procedures.
- Assist sales staff in responding to buyers' questions about warranty procedures.
- Monitor the content of the "Caring for Your Home" section of [Builder] Home Owner Manual to ensure that buyers receive accurate and complete information.

Delivery

- Conduct home owner orientations in a consistent and positive manner.
- Monitor completion of orientation items, ensuring timely response by construction.
- Develop and maintain a report summarizing the number, nature, and completion status of orientation items.

Warranty

- Create procedures for processing routine, emergency, and out-of-warranty items.
- Maintain complete warranty documentation on each home.
- Suggest and organize approved training sessions for warranty staff.
- Schedule and conduct regular performance reviews with each staff person.
- Develop and maintain reports summarizing warranty items.
- Identify recurring warranty items and work with construction and trades personnel to eliminate those that can reasonably be eliminated.
- Meet with newly retained trade contractors to orient them to warranty procedures.
- Evaluate trade contractor warranty performance.
- Work with the trades to improve warranty performance as needed.
- Control costs by efficient use of personnel, appropriate back charges to trades, feedback to construction and purchasing, and common sense application of company warranty guidelines.

Supervisory Responsibilities: Warranty Technicians
                                 Warranty Administrators

Reports to: Chief Service Officer

three aspects of warranty activities also expand, and the company may need more people for repairs and administration or both. If companies add nontraditional services, they need to match staff to the nature of the services they select. Technology that supports all of these efforts can help keep overhead under control while still allowing the company to provide an impressive level of service to home owners.

In-house Warranty Technicians

Some builders take the approach that all warranty work must be performed by trade contractors. This approach often results in repairs taking longer for home owners, and if it leads to disagreements about whether a repair is justifiably the responsibility of a trade, the company may end up paying for this work.

Calculating how many warranty technicians a company needs is a simple matter of dividing the number of homes to be built by the number of homes one technician can care for. A commonly accepted ratio suggests one warranty technician for every 75 homes under warranty. Unfortunately this question cannot be answered that simply. Many factors influence this decision.

Geography. As driving time increases, the number of homes one person can service decreases.

Quality Control. If a company practices strong quality control, that control means fewer warranty items per house, so one technician can service more homes. Builders sometimes react to quality problems by increasing warranty staff. This approach treats the symptom while ignoring the cause.

Size and Design of Product. Small homes usually have fewer warranty items than large homes. Design contributes as well. A small home with a lot of sophisticated details can generate more warranty repairs than a large home with fewer details.

Division of Work. You need to decide which repairs your in-house staff will perform and which ones trade contractors will handle. If in-house personnel perform minor drywall repairs and paint touch-ups, the number of homes each employee can cover will be lower than if trade contractors provide these services.

Courtesy Repairs. Although the investment in courtesy repairs rewards builders with referrals, this commitment does affect workload for warranty technicians.

Competitor's Services. If a competitor announces some terrific new service, pressure builds to respond in kind. Service innovations need not require adding staff, but you must examine workload as part of any new strategy.

Experience and Training. A warranty service veteran can complete work orders faster than a rookie. An experienced technician may be well worth the extra dollars.

Materials and Parts. Having materials or equipment readily available is efficient and increases the number of homes one person can service.

Customer Service History. When customers are satisfied with their homes and their builder, they are less motivated to make excessive demands, so service personnel can handle more homes.

Expectations. Promises made in documents and model homes and by company personnel during sale and construction set buyer expectations. If the promises accurately reflect the company's intentions and abilities, all is well. If the promises exaggerate, you will need more warranty technicians (and possibly legal help).

Seasonal Fluctuations. The number of service technicians required to meet company goals may change as sales fluctuate. Timely adjustments prevent your getting behind on warranty work or wasting dollars through overstaffing.

Suddenly, what began as a simple formula becomes quite complicated. Working backward from your service goals may make your calculations easier. Evaluate your company's warranty service accomplishments by asking the following questions. If the answer to any of these questions is no, you need to analyze why. You may need more people, different people, more training, or better technical support:

* Does the quality of the repairs meet your company guidelines?
* Is repair work completed within desired time limits?
* Is documentation adequate?
* Do other departments receive useful feedback from warranty?
* Are your home owners satisfied?
* Are costs reasonable?

For small companies, the warranty technician job description that appears in Figure 14.4 may be combined with duties for the construction superintendent and the employees hours shared between the two departments. For any size building company, this job description offers a starting point.

Administrative Staff

Nearly all of the factors that affect the number of technical staff also affect administrative staffing levels. If your administrative staffing level fails to meet company goals, study the situation carefully. Eliminate inefficiencies and increase training for current staff before adding more people. If you have technical support, examine how you use it. Administrators sometimes continue to use an old manual system as a comfortable backup and, thus, double their work load. This situation suggests a need for more training to develop confidence in the new system. Once you confirm that the current staff is doing the best possible job and still cannot keep up, temporary help may provide the answer.

A warranty administrator job description appears in Figure 14.5.

Temporary Help

You can bring in temporary help (temp) to support administrative staff or to help field personnel reduce a service backlog. To utilize this approach effectively, keep the following hints in mind:

FIGURE

14.4 Sample Job Description: Warranty Service Technician

Position Title: Warranty Service Technician

General Purpose: Repair items listed on warranty work orders in accordance with [Builder] policies to foster a positive company image.

Duties and Responsibilities:

- Contact home owners to set up repair appointments.
- Order supplies, parts, and materials needed to complete warranty work orders.
- Perform repairs listed on work orders.
- Return completed work orders daily to warranty office, noting any items that require follow up.
- Document any back charges.
- Maintain assigned tools, equipment, and vehicle in good and safe condition at all times.
- Apply all standards and procedures unless authorized by warranty manager to make an exception.
- Perform assigned repairs in model homes.

Supervisory Responsibilities: None

Reports to: Warranty Manager

- Arrange for a suitable work area and see to it that needed supplies are readily accessible before the person arrives.
- Advise the receptionist about the person you expect.
- Introduce the temp to people he or she will be working with or near.
- Give the temp a tour of the office and point out the restrooms, kitchen, emergency exit, and so on.
- Demonstrate how to use the phone system if necessary.
- Explain clearly the work to be done and identify to whom the temp should go with questions.
- As you gain control of the backlog and you see that the temp's work will be coming to a close, notify the temp and his or her agency in writing of the last workday.
- Give the temp agency appropriate feedback.

Nontraditional Service Staffing Numbers

Depending on the particular after-move-in services provided, you may need more support staff. Offering a menu of fee-for-service maintenance requires different staffing from what is needed to send home owner newsletters, develop and present educational workshops, or organize social activities. Ensure that service capacity matches the service commitment. Builders find promising a bit less and doing

FIGURE

14.5   Sample Job Description: Warranty Service Administrator

Position Title: Warranty Service Administrator

General Purpose: Administer all aspects of processing and documenting warranty service requests and other home owner concerns and in accordance with [Builder] policies to foster a positive company image.

Duties and Responsibilities:

Sales and Marketing

- Receive and process requests for work to care for models and show homes.
- Answer routine procedural questions.
- As needed, arrange time for prospects to discuss their warranty questions with the warranty manager.

Delivery

- Receive and file copies of orientation paperwork.
- Receive and forward comments from home owners regarding the status of their orientation items.

Warranty

- Create home owner warranty files.
- Send routine letters and reminders to home owners at appropriate times.
- Respond to home owner contacts: phone, fax, e-mail, letter, or in-person visits to the office.
- Input warranty items into the computer.
- Issue and distribute approved work orders.
- Monitor completion of work orders and update computer records.
- Prepare and distribute reports summarizing warranty work.
- Create correspondence to home owners and trades.
- Assist in-house and trade personnel in setting up appointments for inspections and repairs.
- Initiate follow-up calls to home owners and trades.
- Document after-hours emergencies and follow up to ensure completion of needed collateral work.

Supervisory Responsibilities: None

Reports to: Warranty Manager

it well is more effective than promising a lot and being unable to deliver it in a world-class manner.

Common sense suggests that providing maintenance services to home owners would involve the builder's existing warranty staff. They are familiar with the product, the home owners, and even the streets in the communities. Coordinating service extras with regular warranty visits is efficient and less disruptive to the home owners' schedules. However, you must take into account the time demands of these extras when you decide on the number of personnel needed.

A Centralized or Decentralized Warranty Office

Industry veterans vigorously debate which works "best"—centralized or decentralized warranty service. In truth, neither is best. As with staff levels and structure, search for the method that fits your situation most appropriately, and keep in mind that what works well this year may need to be revamped next year. Consider these factors:

Geography. Even short distances in heavy traffic can be detrimental to service. Tolls and road conditions may enter into your thinking, as well as where your employees live relative to the work site(s).

Control. Will your management team be comfortable without daily observation of the activities of field service personnel? If you trust the people you have hired to have the integrity to put in a full day's work and the knowledge to do so effectively, this control should not be an issue.

Physical Facilities. Do you have a suitable location for an on-site warranty office? A desk, phone, fax, possibly computer, files, and so on are important support tools for prompt warranty work.

Documentation. How will you manage documentation? Consistency can quickly become the victim of decentralizing unless you have clear procedures in place and all personnel follow them. Training the warranty team on an appropriate computerized system can ensure consistency.

Personnel. Does your company employ sufficient numbers of skilled personnel to set up separate warranty offices? How will those personnel stay connected to the rest of the company: weekly in-office staff meeting, e-mail, phone? Will your department head make frequent field visits?

Closeout. How long will you need the satellite office? Does the time frame make sense when balanced against the set-up cost?

The Service Attitude

Whether or not your company provides traditional warranty service or goes beyond it to provide

Builders can make a good argument for satellite warranty offices for mid- to large-size companies. Home owners appreciate having someone available or near-by. Familiarity with the community can generate creative ideas, help the warranty technicians stay in touch with local issues and events, and help them avoid missteps (such as planning the Autumn Service Sweep for the same week school starts).

When preparing to close a satellite office, begin with a letter to the affected home owners. Explain clearly how the company will manage warranty service after you move the local warranty office. Ensure that these home owners continue to receive prompt responses and that follow-up efforts provide promised repairs.

nontraditional services, the warranty team involved are critical to successful after-move-in service. No system or set of procedures is so clever that it is self-propelled. The right people—those with the right attitudes, knowledge, and skills—are critical to success.

Think Long-Term

Turnover is costly. It negates the benefits of training and leaves you starting over again and again. Select staff carefully using a long-term view. Determine what skills your warranty office needs now and what it will need in the future. Hire people who can grow with your company.

Warranty Staff

Warranty personnel should excel as problem solvers. The work requires common sense and sound judgment. They must be able to diagnose problems and propose solutions consistent with company policies that will satisfy home owners. They must remain calm and composed, effective in emergencies, and controlled under stress. A healthy ego, a good sense of humor, and the ability to receive criticism without taking it personally enable an employee to remain friendly and courteous while being firm.

A construction background is clearly desirable. Familiarity with the fundamentals of scheduling, construction techniques, and terminology all help get the job done. If you cannot find a person with the exact combination of desired traits look for someone with people skills and add the construction training. With training in the field, planned field experience, a bright, motivated individual could learn quickly.

Hiring warranty personnel with the promise that after six months in warranty they will be promoted to the position they really wanted—usually superintendent—can lead to problems. Although superintendents benefit from seeing the end result of their work, when warranty service is a stepping stone to other company positions, turnover can become continuous and interest in warranty work diminishes. Home owners repeatedly see new names and faces; as a result they need to keep repeating their house histories and service particulars. One rotating internship position among a larger warranty staff causes fewer problems, but core personnel need to be committed to warranty work.

Staffing for Nontraditional Services

Those staff members involved in planning and implementing nontraditional after-move-in services need enthusiasm, creativity, and strong organizational and communication skills. They also should be genuinely interested in home owners. Look for friendly, energetic people who can think outside the traditional warranty box and still stay within a reasonable budget.

Training and Development

Staff members can only serve your home owners to the extent they know how. Provide comprehensive training of staff and include an annual needs assessment, budget, and schedule. Over time your company should develop a curriculum with training plans for all functions that covers their particular technical expertise, customer service, communication, and career skills such as computer use and time management. Small-volume builders could take advantage of local community college classes, generic seminars, regional and national conferences, and the other resources listed, but they probably would not have an extensive in-house training program. No matter what your volume, begin with a thorough new employee orientation and continue with some of the training suggestions listed in Chapters 2 through 10.

New Employee Orientation

A thorough orientation helps new employees become effective sooner and creates the best chance of their long-term success with your company. Plan for new employee orientation to take several weeks: Devote the employee's entire first day to it and provide shorter sessions once a week thereafter.

* Begin with an overview of the orientation schedule and suggest that the new employee keep a list of questions for periodic discussion with the supervisor.
* Review the company's background: how it was started, significant achievements, current status, and future vision.
* Explain how feedback from home buyers and the trades is reviewed and used. Look at recent levels of satisfaction and explain current action items.
* Review the new employee's job description in detail, discuss where the job fits in the company organizational chart, and review responsibilities and success criteria.
* Explain the company's Universal Service Guidelines
* If your company has an employee manual, review its contents and have the employee fill out the necessary paperwork for taxes, insurance, and so on. In large organizations, a human resource manager may perform this step.
* Tour the office facility and introduce the newcomer to other employees.
* Schedule follow-up time in each department or function so that the newcomer has an opportunity to learn how the company's work gets done. Figure 14.6 is a sample procedure review worksheet that you can use for initial employee orientation as well as for ongoing cross training.
* The newcomer should visit all the communities in which you build and tour model homes, not once, but several times, so provide a map and driving directions if necessary
* Staff should learn how your company puts a home together, first on paper and then in reality so they can answer home owners' questions on the spot. Plan tours of homes at various stages of construction so that the new staff person can observe construction sequence and techniques.
* Component by component—beginning with standard items and working through options and common upgrades—warranty staff should learn all they can about each material or product that your company includes in its homes. Anticipate what home owners will need to know. Assign a product to each

FIGURE

14.6 Procedure Review

[Logo]   Procedure Review

Procedure ____Selection Appointment_____

Documentation. Gather copies of the documents related to this procedure, list them below, and check each one off as you read it.

☑ ___Standard feature list_____

☑ ___Displayed selections: model info sheet_____

☑ ___Change order policy and sample change order_____

☑ ___Home owner guide section on selections_____

☑ ___Selection sheets (4 pages, minimum)_____

☐ _____

Interviews. Interview one or more individuals who work with clients to perform this procedure. Attach notes as needed.

Date	Name
✓	Charmel Rhineler, Manager
✓	Mary Ellison, Coordinator
✓	Ryan Yardley, Coordinator

Observations. Observe one or more appointments with clients to see a real example of this procedure. Attach notes as needed.

Date	Name
9/11/20–	Mr. & Mrs. Bryson
10/7/20–	Ms. Ericson
11/9/20–	Mr. & Mrs. Hernandez

FIGURE
14.6 *Continued*

Conclusions. Record your thoughts and conclusions regarding this <procedure>. What are the strengths of this procedure, and where are the opportunities to improve this part of the new home process in your company? Attach your notes below or following this page.

Preparing buyers for the number of choices they will make is vital; the greeting from reception sets a nice tone.

Meetings seem to take a long time. The coordinators get worried about messages piling up. (Can the receptionist help with this?)

Construction might benefit from working with the coordinators to help them better prepare buyers for the next meeting (pre-construction). They do not seem clear on how those work.

person to study and report on using the component review (Figure 14.8). Ask warranty staff to review a section of your home owner guide; assign a topic or two prior to each staff meeting. Discuss the material. When you reach the end, start over.

Communication and Cross Training

Regular communication is critical to effective daily functioning even if each department consists of only one or two people. Friction among sales, construction, and warranty usually results from lack of understanding of each other's goals, methods, and challenges. Communication and understanding can reduce the "natural enemy syndrome." Understanding and mutual respect translate into supportive daily behaviors such as the superintendent reminding a trade of the importance of completing warranty items or the warranty manager talking with a sales prospect to answer technical questions.

Cross-training activities might include job shadowing, interviewing those who perform a task or observing them perform it, reviewing procedures and reports, and reading company documents. Occasional meetings among the warranty, sales, and construction departments will keep the staffs of those departments or functions updated on changes. Regular communication is critical to

FIGURE
14.7 Product Profile

[logo]

Product Profile

Community	Bailey's Cove	
Location	Bailey CO	
Directions	2 miles east of Hwy X 149 and turn left @ gas station,	
	1/2 mile to community on right	

On-site personnel	Sales	Construction
	Cassandra Marx	Eric Freeman
Name(s)	Elizabeth Rylie	Tom Belmont
Phone	555-555-3009	
Cell phone	Liz: 555-707-1419	Eric: 555-440-1338
Fax	555-555-3009	
E-mail	lizR@aol.com	tomb@builder.com

Floor plan*	Base price	Sq. ft.	Comments
Trinity Bay	450,600	2480	4 BR, nice family room
Port View	462,500	2694	Large kitchen, great DR
Sunrise	497,300	2800	5 BR or loft option
Bay Port	510,800	3150	Great home office—decor is
			dark and dramatic

*Attach 8.5 × 11-inch floor plans if available

Community Data

☑ Number of homes at completion ___208___
☑ Date community opened ___10/1/20–___
☑ Anticipated closeout date ___10/30/20–___
☑ Merchandising (brochures, feature lists, etc.)
☑ Purchase agreement and addenda
☐ Option list ___Unavailable—being updated___

☑ Selection form
☑ Change order policy and change order form
☑ Home owner manual
☑ Home owner association documents ___NA___

FIGURE

14.7 *Continued*

Tour Models and Show Homes

Date _____4/1/20–_____   Date _____

Date _____10/91/20–_____   Date _____

Date _____5/16/20–_____   Date _____

---

effective daily functioning even if each department consists of only one or two people.

Performance Reviews

Engage in ongoing dialog with employees about their performance. Set a timetable for performance reviews and follow it faithfully. Constructive guidance is the first step in improving employee performance. Select one or two issues to target for improvement and present them in a positive tone. Include an opportunity for the employee to make observations and to participate in setting goals.

Compensation

Finding the right people to work with your customers on their warranty requests can be difficult and time consuming. Orientation and training costs add to your investment. Turnover costs money and can set the builder's service program back by months, perhaps even years. Competitive compensation can help to prevent that loss.

Your company is legally obligated to provide quick, effective warranty repairs. By bringing enthusiasm to the task, hiring good people, training them well, and paying them generously, you can generate the most powerful form of advertising available: the recommendations of satisfied customers. Thus, improving service brings you word-of-mouth advertising—the best kind—while increasing paid advertising to compensate for a poor service reputation spends hard-earned profits inefficiently.

Incentives

For bonus programs to improve employee performance while avoiding ill-effects, they should

- reflect the company's culture and long-term goals
- involve the participants in creating bonus programs and the criteria for awarding bonuses

FIGURE
14.8 Component Review

[Logo]   Component Review

Component or System

Laminated Countertop

Standard Product

Selections offers several brands—all care is the same.

Options/Upgrades. List them.

NA

Function. List points to discuss, features to demonstrate.

Importance of caulking top to backsplash to keep water
from penetrating.

Use and Care. Describe cleaning, routine maintenance, and troubleshooting techniques.

Avoid abrasive cleanser, cigarettes set on edge, hot iron, use of
knives and other sharp objects.

Builder Limited Warranty. List what your company repairs and what limitations apply.

Highly polished finishes are excluded from warranty and
selections recommends against them; buyer would sign waiver.

Manufacturer Warranty. Describe the warranty coverage the manufacturer provides.

Laminate should adhere—no bubbles.
Seams are to be sealed and should be smooth.

Experience. Provide feedback from home owners.

Car polish can be used to produce a nice finish.

Resources. List names, addresses, phone numbers, e-mail address, and Web site if any.

Kate Benson at Classy Counters very
knowledgeable and willing to talk to home owners.
Cell: 555-440-9001

* use objectives and criteria that are attainable and still challenging
* be equitable; reward people for activities over which they have control
* include rewards that are interesting, relevant, and worthwhile to the employees earning them
* be of specific, preferably short, duration (no more than six months per program)
* compare performance to objective criteria, rather than to that of other individuals or teams
* occasionally include programs for trades
* encourage team effort and cooperation
* supplement, not replace, respectable pay

Many well-intended bonus programs suffer from unanticipated consequences. Although ill-structured bonus programs may produce results at first, be wary of the following signs that your bonus program needs an overhaul:

* Bonuses become a way of life. Rewards are taken for granted and expected in return for routine work, rather than being a stimulus for extra effort.
* Those who do not "make bonus" become discouraged and may give up.
* Employees who contribute to successful service but who are omitted from the bonus system become discouraged or resentful. Customer satisfaction is a team effort.
* Bonuses send a message that the job being rewarded has little intrinsic value. The implication is that no one would do the job for its own sake.
* Bonuses unintentionally reward the wrong behaviors. Sometimes employees feel manipulated and manipulate back. They work the system to "win" against a bonus program they may believe is unfair.
* Bonuses cause company personnel to pull in opposite directions, as in a program that rewards the vice president for meeting a quota of closings, superintendents for bringing homes in under budget, and warranty staff for home owners' satisfaction. You will need to decide before priorities come into conflict, whose authority (and priority) will rule.

Perhaps most important, bonus programs should exist only with honest communication with the frontline about what obstacles exist to doing their best. When a company's key issues are candidly addressed by a management that stays in touch with employees' daily realities, bonus programs have their best chance to be effective.

Recognition

From a simple thank you to more formal acknowledgments such as wall plaques, employees (and trade contractors) appreciate the fact that you noticed their efforts. What your company measures and rewards will typically increase. Look for ways to recognize service excellence to encourage more of the same. Company newsletters, the bulletin board in the snack room, announcements in pay envelopes, press releases, and ceremonies at all-company meetings offer a variety of methods. A pat on the back—especially in front of others—is a powerful motivator.

Retaining Service Talent

Talented people naturally want to work for companies of which they can be proud and that respect their efforts. Therefore, your company's reputation affects employee recruitment as well as negotiations with vendors and trades. Employee loyalty evolves from employees seeing that they are making tangible contributions to company success and that the company recognizes and appreciates those contributions. Employees' enthusiastic attention to home owners reflects their loyalty. Treated well, home owners in turn may mirror that loyalty—and that reflects well on everyone.

Service: Your Competitive Edge

From legal and insurance necessity to satisfaction rankings, the home buyer today brings more forces to bear on builders for better homes and better service. Smart builders have responded and in many cases have successfully defused issues and won hearts and minds. A remarkable amount of progress has occurred in the service aspect of the new home experience. And builders have more to do.

The policies and procedures of companies that have been wildly successful in achieving customer satisfaction are no secret. Attend any national or regional conference and you hear discussion of service methods and techniques. Pick up any industry periodical and you find graphs and charts, interviews, and reports on how various companies won awards for their customer satisfaction levels. Clearly, none of these firms has discovered or invented a silver bullet, but they do have the following items in common:

- They have invested time and money, thought and planning, training and cross training.
- They do the basics, and they do them well: communication, quality inspections, more communications, training and more training.
- Through trial and error they have identified the approach that works well with their customer bases.

At times this work is tedious. You must address the fundamentals effectively. You must excel in the less interesting routine of day-to-day basics before you can move on to exciting new ideas.

Few Achieve Perfection

Some days, improvement in customer satisfaction appears to be an impossible goal; the home buyers seem too unpredictable and the tasks too arduous. One tough home buyer can distract both frontline employees and management from their larger goal of satisfying home buyers, and that distraction allows discouragement to spread. Yet persistence pays off.

One Size Does *Not* Fit All

Although no two programs are exactly alike, nearly all home builders now understand phrases such as *touch points* and *managing the experience*. However, each company's personality and capacity creates a unique environment. Company owners, even in the same geographic area and at the same price point, seldom see their markets exactly the same way. Each manager thinks of different service components, unusual twists on classic approaches, or completely innovative touches.

Your search is for the combination that wins you more sales. No universally right program exists. That fact can both comfort and frustrate you. No one can hand you a clear and proven formula. But, simultaneously, you have the opportunity to create your own approach without being tied to strict requirements. The potential is almost limitless, and creative professionals challenge themselves to new levels of customer empathy and sensitivity, as they search for exciting touches that make potential customers buy now and refer their family and friends later.

Expecting to hit the target with the first effort or with every new element you add to service would be naive. **Experience will narrow the list of potential methods you use, as you discard failures and each success builds momentum.**

Signs of Progress

As customer satisfaction grows, so does positive word-of-mouth marketing. What may have seemed for months (or years) like a thankless chore begins to provide a return on investment. Encouraged, companies rethink, reform, and retool for still better relations with home owners.

Builders now seek ideas and suggestions from those involved with the building process. Customer feedback, a process still in its infancy, has become a common management tool. Trade partnering programs abound, and these programs tie the people who contribute in the most tangible way to the goals of satisfaction and increased sales. Company owners encourage staff members to suggest ideas for improving customer service. They also dramatically increase training and cross training of their personnel because they see the value of putting competent, caring staff to work for home buyers.

An Ongoing Challenge

Much as everyone one involved would like to reach an ultimate service destination, none exists. Complacency has no place in this effort. When you realize you have implemented a winning combination of ideas, well-executed by true professionals, your competition will take note and emulate your achievement. Meanwhile, the home buyers themselves become accustomed to the new level of service and begin to wonder what will be next. Builders have just begun exploring this new service frontier, and you are one of the pioneers.

Selected
Bibliography

Beckwith, Harry. *What Clients Love: A Field Guide to Growing Your Business.* New York: Warner Books, 2003.

Berry, Leonard. *Discovering the Soul of Service.* New York: The Free Press, 1999.

Blanchard, Ken, and Sheldon M. Bowles. *Raving Fans.* New York: William Morrow & Co., 1993.

Carbone, Lewis. *Clued In: How to Keep Customers Coming Back Again and Again.* New York: Financial Times/ Prentice Hall, 2004.

Carlzon, Jan. *Moments of Truth.* New York: HarperCollins, 1989.

Carnegie, Dale. *How to Win Friends and Influence People.* New York: Pocket Books, 1998.

Gittell, Jody Hoffer. *The Southwest Airlines Way.* New York: McGraw-Hill, 2003.

Gross, T. Scott. *Why Service Stinks . . . and Exactly What to Do about It!* Chicago: Dearborn Trade, 2003.

Jaffe, David, and David Crump. *Contracts and Liability*, 5th ed. Washington, D.C.: BuilderBooks, National Association of Home Builders, 2004.

MacKenzie, Alec. *The Time Trap*, 3rd ed. New York: Amacom, 1997.

Morgenstern, Julie. *Time Management from the Inside Out.* New York: Owl Books, Henry Holt & Co., 2004.

Parisi-Carew, Eunice. *The One Minute Manager Builds High Performing Teams.* New York: William Morrow and Co., Inc., 2000.

Popcorn, Faith, and Lys Marigold. *Clicking: 17 Trends That Drive Your Business—And Your Life.* New York: Harper Collins, 1996.

_____. *EVEolution: Understanding Woman—Eight Essential Truths That Work in Your Business and Your Life.* New York: Hyperion, 2001.

Smith, Carol. *Building Your Home: An Insider's Guide.* Washington, D.C.: BuilderBooks, National Association of Home Builders, 2005.

_____. *Customer Service for Home Builders.* Washington, D.C.: Builder Books, National Association of Home Builders, 2003.

_____. *Dear Homeowner: A Book of Customer Service Letters.* Washington, D.C.: BuilderBooks, National Association of Home Builders, 2000.

_____. *Homeowner Manual: A Template for Home Builders.* Washington, D.C.: BuilderBooks, National Association of Home Builders, 2001.

_____. *Meetings with Clients: A Self-Study Manual for a Builder's Frontline Personnel.* Washington, D.C.: BuilderBooks, National Association of Home Builders, 2002.

Reichheld, Fred. *The Ultimate Question: For Unlocking the Door to Good Profits and True Growth*. New York: Harvard Business School Press, 2006.

Van Bennekom, Frederick C. *Customer Surveying: A Guidebook for Service Managers*. Bolton, MA: Customer Service Press, 2002.

Zaltman, Gerald. *How Customers Think: Essential Insights Into the Mind of the Market*. New York: Harvard Business School Press, 2003.

Index

meeting matrix, 5, *6–7*, 8–12
meetings. *See also* Universal Service
    Guidelines
    staff, 76, 94, 156, 192–93, *194*
    with home buyers/owners, 5, *6–7, 8,*
        8–12, *15, 16,* 46, 48, 63
    community team, 14, 17–18, 195
missed appointments, 134
monitoring, 129–38
multiple-trade repairs, 116–17, *117, 118,*
    128, 133

N

Nassof, Russell, 107–108
neighbor disagreements, 121–22, *123*
nonwarranty personnel, receiving warranty
    requests, 72, 74
Notice and Right-to-Repair Laws, xv, 129,
    197

O

objectives, after-move-in service, 198
on-call duty for staff, 74
one-year limited new home warranty agree-
    ment, sample, *20–22*
one-year and two-year materials and work-
    manship warranties, 27
ordering materials, 99
orientation, 54–55, 81, 83,160–65 *161, 163,*
    233, *234–35, 236,* 237

P

payment, 132, 156, 162
phone log, 72, *73*
physical judgment, 83–84
policies and procedures, 57, 160, 241
predictable situations, 149
procedure review, 233, *234–35*
product profile, *236–37*
projects, manage, 185
property management services, 207–208
purchase programs, for employees and trade
    contractors, 214

Q

quality, 79, 102, 135, 155, 225
quality standards. *See* guidelines

quality control, 72, 227
quiz, 28, *29*

R

Reicheld, Frederick C., 61
referrals, 197
    appreciation, 214
    from home owner, 213–14,
    to builder's attorney, 147, *148*
    rehersals and role plays, 77, 94, 166, 211,
        213–14
    training employees to ask for, 63, 213–14
    to warranty insurance company, 35, 144,
        146, *147*
repairs, 95–104
    access for, 132
    check in lieu of, 134, *135*
    courtesy, 227
    damage caused during, 103
    defective material, 105, 111, *112*
    expectations, 92
    follow up after completion, 136–38
    guidelines for, 31–52, 166
        *Residential Construction Performance
            Guidelines,* 3rd ed., 31
        Universal Service Guidelines, 218–219,
            220–223
    home owner signature on work order,
        103
    inappropriate conditions for, 102
    industry standards, 50
    method of, 135
    multiple-trade, 116–17, *117, 118,* 128,
        133
    notes about, 103
    obligation for, 24
    one-trip, 101–102
    outside scope of limited warranty, 143,
        145
    payment for, 132, 156, 162
    quality of, 79, 102, 135–36, 155, 223
    reponse time for, 127, 159–60
    roof leaks, 108–109, *109*
    tracking, 80,
    warranty appointment guidelines,
        97–102, *100–101,* 160, 166
    wet basement, 109, *110*
    work date approach, 98, 99
    work order. *See* work order.
report. *See* warranty report.
reputation, 1, 86

The National Association of Home Builders is a Washington-based trade association representing more than 235,000 members involved in home building, remodeling, multifamily construction, property management, trade contracting, design, housing finance, building product manufacturing, and other aspects of residential and light commercial construction. Known as "the voice of the housing industry," NAHB is affiliated with more than 800 state and local home builders associations around the country. NAHB's builder members construct about 80 percent of all new residential units, supporting one of the largest engines of economic growth in the country: housing.

 Join the National Association of Home Builders by joining your local home builders association. Visit www.nahb.org/join or call 800-368-5242, x0, for information on state and local associations near you. Great member benefits include:

- Access to the **National Housing Resource Center** and its collection of electronic databases, books, journals, videos, and CDs. Call 800-368-5254, x8296 or e-mail nhrc@nahb.org
- **Nation's Building News**, the weekly e-newsletter containing industry news. Visit www.nahb.org/nbn
- **Extended access to www.nahb.org** when members log in. Visit www.nahb.org/login
- **Business Management Tools** for members only that are designed to help you improve strategic planning, time management, information technology, customer service, and other ways to increase profits through effective business management. Visit www.nahb.org/biztools
- **Council membership**:
  - **Building Systems Council**: www.nahb.org/buildingsystems
  - **Commercial Builders Council**: www.nahb.org/commercial
  - **Building Systems Council's Concrete Home Building Council**: www.nahb.org/concrete
  - **Multifamily Council**: www.nahb.org/multifamily
  - **National Sales & Marketing Council**: www.nahb.org/nsmc
  - **Remodelors™ Council**: www.nahb.org/remodelors
  - **Women's Council**: www.nahb.org/womens
  - **50+ Housing Council**: www.nahb.org/50plus

 BuilderBooks, the book publishing arm of NAHB, publishes inspirational and educational products for the housing industry and offers a variety of books, software, brochures, and more in English and Spanish. Visit www.BuilderBooks.com or call 800-223-2665. NAHB members save at least 10% on every book.

 BuilderBooks Digital Delivery offers over 30 publications, forms, contracts, and checklists that are instantly delivered in electronic format to your desktop. Visit www.BuilderBooks.com and click on Digital Delivery.

 The **Member Advantage Program** offers NAHB members discounts on products and services such as computers, automobiles, payroll services, and much more. Keep more of your hard-earned revenue by cashing in on the savings today. Visit www.nahb.org/ma for a comprehensive overview of all available programs.